EAST OF ENGLAND STEAM
1948-1963

PETER TUFFREY

GREAT NORTHERN

ACKNOWLEDGEMENTS

I am grateful for the help received from the following people: Roger Arnold, Paul Chancellor, David Burrill, John Chalcraft, D.J. Dippie, Hugh Parkin, Bill Reed.

Gratitude should also be expressed to my son Tristram for his general help and encouragement throughout the course of the project.

Great Northern Books
PO Box 1380, Bradford, BD5 5FB
www.greatnorthernbooks.co.uk

© Peter Tuffrey 2023

Every effort has been made to acknowledge correctly and contact the copyright holders of material in this book. Great Northern Books Ltd apologises for any unintentional errors or omissions, which should be notified to the publisher.

All rights reserved. No part of this book may be reproduced in any form or by any means without permission in writing from the publisher, except by a reviewer who may quote brief passages in a review.

ISBN: 978-1-914227-46-2

Design and layout: David Burrill

CIP Data
A catalogue for this book is available from the British Library

INTRODUCTION

For several reasons, the grouping of areas within England has been beneficial for the Government and population. Whilst the country had been arranged in such a way briefly during the Civil War, politicians only grasped the idea firmly in the early 20th century. They planned to devolve some aspects of governance to local levels where greater responsibility and accountability were possible. Nine 'Standard Regions' were created after the Second World War, then in 1994 the ten in use at the present time were formed. These were the North East, North West, Yorkshire and the Humber, East Midlands, West Midlands, South East, South West, London and the East of England.

The last-mentioned covers over 7,000 square miles, being the second largest region by area, and at present has over six million inhabitants. The counties comprise: Bedfordshire; Cambridgeshire; Essex; Hertfordshire; Norfolk; Suffolk. Huntingdonshire has a historical connection with the area also. Some of the important towns and cities incorporated in the East of England are: Norwich; Ipswich; Colchester; Cambridge; Peterborough; Bedford; Yarmouth; Ely; King's Lynn; Lowestoft; March.

The East of England also offers an interesting area of study for the railways. Though many small companies built a number of the main routes in sections, the region was dominated by the Great Eastern Railway from the 1860s to the Grouping of 1923 when amalgamated into the London & North Eastern Railway. Whilst there was a transfer of freight to the west (Peterborough, Hitchin and Bedford), there was little intrusion by other companies, except for the Midland & Great Northern Joint Railway. The line reached right across the northern half of the area to connect the Midlands with Norwich and Yarmouth/Lowestoft on the East Coast.

As with other areas of the country, early railway schemes were developed in the 1820s. One was formulated to run from Bishop's Stortford to the River Cam at Cambridge and another would have seen lines laid from London to Cambridge. Receiving parliamentary approval in the mid-1830s was the Eastern Counties Railway which planned to run from London to Colchester, Ipswich, Norwich and Yarmouth. Around three years elapsed from authorisation to the first section from Mile End to Romford being ready, owing to challenges in construction around Stratford. Further troubles in the form of high costs and lack of capital almost saw the line abandoned soon after, yet this was averted and progress was made to Brentwood in the following year. The problems were far from over, however, and Colchester was not reached until 31st March 1843. Up to this point, the line had cost the same amount as that quoted for the whole of the route to Norwich.

With only part of the original scheme fulfilled, and little chance of further progress, the people of Norfolk were aggrieved enough to start a new venture to join Norwich and Yarmouth. The 20 miles of railway were laid in just a year, being ready in April 1844. Similarly, in Ipswich the population felt isolated and envious of developments to the north and west. Therefore, the Eastern Union Railway was founded to join the line to London at Colchester. The first train ran from Ipswich to Colchester on 11th June 1846.

Another ambitious project mooted early in the area's railway history was a line from East London to York, via Cambridge. The Northern & Eastern Railway was formed in the mid-1830s for this task, yet Parliament only allowed a section of this from Stratford to Cambridge. As with the Eastern Counties Railway, financial trouble limited even this project and by 1840 the line ran as far as Broxbourne. The company inched towards Cambridge and in 1843 boasted a route to Bishop's Stortford. At this time, the ECR arrived at an agreement with the N&ER to take over the route and construct the remainder. Whereas the ECR struggled with their original line, the progress on this, and two authorised extensions to Peterborough via March and Ely and to Brandon, was relatively rapid. Travel by rail between London and Norwich was finally possible in 1845 thanks to the connection at Brandon with the Norfolk Railway.

As several lines were developing to the south of the region, people in the north quickly sought to connect their areas with the new routes. The Lynn & Ely Railway branched southward from King's Lynn to join the ECR at Ely in 1847, as well as another short railway to meet the company at Wisbech. A year later, the Lynn & Dereham Railway met the Norfolk Railway which laid a line northward from Wymondham. The Newmarket & Chesterford Railway also joined the ECR south of Cambridge for horse racing traffic, yet the company fell into financial problems and the original rails were lifted and re-laid to reach the main line north of Cambridge.

Following the pattern set elsewhere in the country, the dominant company in the area swallowed the remaining smaller companies. In East England, the Eastern Counties Railway took over the Eastern Union Railway, Norfolk Railway, East Anglian Railway and Newmarket Railway, creating the Great Eastern Railway in 1862. The new company was ambitious to tap the coal reserves of Yorkshire and the North with a railway, yet opposition was found from the Great Northern Railway. This company had evolved from the original idea of the Northern & Eastern to connect York via Cambridge but the GNR ran via Peterborough, Grantham and Doncaster, while also looping through Lincolnshire via Spalding, Boston and Lincoln. The GER and GNR vacillated over a joint proposal through to 1879 when finally setting aside their differences. This route was laid from March to Spalding, Sleaford and Lincoln to Doncaster, opening in the mid-1880s.

The GER inherited the precarious finances of the

constituents. This left several independent companies free to lay lines and compete with the larger company. The area west of King's Lynn, connecting Spalding and Peterborough, was captured by small companies and to the east the Lynn & Fakenham Railway was built, along with several others. The GNR and Midland Railways soon joined forces to amalgamate these, owing to the closeness of their own enterprises. Those in the east thought joining forces was advisable and in 1883 the Eastern & Midlands Railway began operations. Running into accounting problems at the end of the decade, the GNR and MR stepped in to take over and from 1893 operated the Midland & Great Northern Joint Railway. From Little Bytham to Peterborough, two sections merged at Sutton Bridge to run into King's Lynn. The line then ran to Melton Constable, splitting there, ultimately, to Cromer, Norwich and Yarmouth.

To work trains in the area, the GER developed several locomotive classes that were robust and reliable. At the start of the 20th century, James Holden produced the first S46 (LNER D14) Class 'Claud Hamilton' 4-4-0 for East Anglian expresses and continental boat trains. S.D. Holden improved on his father's design for the S69 (LNER B12) Class 4-6-0 which were necessary owing to increased train weights before the First World War. For freight traffic, the company relied on 0-6-0s of various designs from James and S.D. Holden, as well as the latter's successor A.J. Hill. These were mainly classified J15, J17, J19 and J20 by the London & North Eastern Railway after Grouping in 1923. The GER had an intense suburban service running around London's East End and into Essex. For this, the GER stuck with the 2-4-2T design for a number of years. Not until A.J. Hill's appointment did an 0-6-2T appear which was in line with contemporary practice and the design was perpetuated in numbers throughout the 1920s.

Many of the aforementioned engines were kept in traffic after the formation of the London & North Eastern Railway. Particularly as permanent way restrictions limited the motive power available for work in the area for many years. Gresley had trouble designing the B12 replacement within these constraints and the B17 construction process was protracted as a result. On the main lines and M&GNJR, the Gresley K3 Class 2-6-0s were employed successfully on both passenger and freight duties and the J39 0-6-0 supplemented the GER types on goods trains.

The Second World War kept a number of classes in use beyond their lifespan. Several 2-4-2Ts found further use on local branch trains away from London, whilst E4 Class 2-4-0s kept working in the Cambridge area. For similar duties, 'foreign' locomotives were brought into the area, such as H.A. Ivatt's C12 Class 4-4-2T and Wilson Worsdell's G5 Class. Replacements gradually appeared, firstly with Edward Thompson's B1 Class 4-6-0 and L1 Class 2-6-4T, then British Railways' Standard Class 7 'Britannia' Pacifics. The latter eclipsed the B17 and B1 on the principal expresses and boat trains throughout the 1950s. For M&GNJR services, H.G. Ivatt's London Midland & Scottish Railway 4MT 2-6-0 design was adopted to take over from older classes until the line's closure in 1959.

Part of the Great Eastern Main Line to Brentwood was electrified after the war and the East of England converted to diesel locomotives quicker than other areas of the country. Many of the local lines were also the victims of the closures that took place in the mid-1960s. *East of England Steam 1948-1963* captures the final years of steam traction in the region when the railways were intact and enjoying their final years in an almost forgotten time of English history.

Peter Tuffrey
Doncaster, January 2023

Above ALDEBURGH STATION – NO. 65447
During April 1956, Worsdell J15 Class 0-6-0 no. 65447 has a Saxmundham train at Aldeburgh station. The latter was the terminus of a branch from Saxmundham which was on the main line to Lowestoft. Photograph courtesy Rail-Online.

Below ALDEBURGH STATION – NO. 67230
The 10.25 train to Saxmundham waits to depart from Aldeburgh station behind Holden F6 Class 2-4-2T no. 67230. Pictured on 7th April 1956, the locomotive transferred from Lowestoft to Stratford at the end of the year. Photograph courtesy Rail-Online.

Above BARNWELL JUNCTION STATION – NO. 62559
To the north east of Cambridge, Barnwell was the location for the divergence of the Mildenhall branch. Constructed by the Great Eastern Railway at the end of the 19th century, the line was originally intended to be an alternative to the Norwich route, which ran via Lakenheath, and instead of reaching Thetford, terminated at Mildenhall. The branch was completed in 1884/1885 and Barnwell station was sited just off the main line to King's Lynn, being just out of frame on the right in this image from April 1955. Holden D16/3 Class 4-4-0 no. 62559 is running southward towards Cambridge with a train from King's Lynn. Photograph courtesy Rail-Online.

Below AYLSHAM SOUTH STATION – NO. 65566

James Holden introduced the F48 Class 0-6-0 in 1900 for heavy goods services. Sixty of these were built with a round-top firebox before Holden switched preference to the Belpaire-type which had a flat top. The latter were classified G58 (LNER J17) and a total of 30 appeared between 1905-1911. No. 65566 (amongst the first batch of G58 engines built) was completed at Stratford in May 1905. The locomotive has a goods train bound for Reepham at Aylsham South station on 29th August 1958. This had originated at Norwich, where no. 65566 was based between June 1955 and July 1960 when condemned for scrap. Photograph by Hugh Ballantyne courtesy Rail Photoprints.

Above BARNWELL STATION – NO. 45302
On 15th June 1963, just under a year remained for Barnwell station as closure of the Northampton-Peterborough route occurred in May of the following year. With the 18.00 westbound local train here is Stanier Class 5 4-6-0 no. 45302. The locomotive was amongst the pre-War class members, being built by Armstrong Whitworth & Co. during January 1937. In traffic for 30 years, no. 45302 was condemned at Stoke shed. Photograph by Tony Cooke courtesy Colour-Rail.

Opposite above ARDLEIGH STATION – NO. 70008
The 'East Anglian' express was introduced by the London & North Eastern Railway to be an 'exclusive' service between Liverpool Street station, Ipswich and Norwich. Operating from 1937-1939, the train returned after the Second World War and was joined by the 'Norfolkman'. The latter left the capital for East Anglia in the early morning and returned in the early evening, whereas the 'East Anglian' ran the opposite schedule. Initially, the trains were headed by Thompson B1 Class 4-6-0s before the introduction of British Railways' Standard Class 7 'Britannia' Pacific in the early 1950s. No. 70008 *Black Prince* has the westbound 'Norfolkman' at Ardleigh station in the mid-1950s. Photograph courtesy Rail-Online.

Opposite below BARNWELL STATION – NO. 45113
A wreath adorns the front of Stanier Class 5 4-6-0 no. 45113 to mark the closure of the line from Northampton to Peterborough on 4th May 1964. The London & Birmingham Railway passed to the west of Northampton before a branch to the town was promoted in the early 1840s. At the same time, the L&BR thought an extension to Peterborough was advantageous as the city had not yet been reached by the Eastern Counties or Great Northern Railways. The line opened throughout in June 1845, with Barnwell – west of Peterborough – serving passengers from this time to closure. Photograph by Tony Cooke courtesy Colour-Rail.

BARTLOW STATION – NO. 67322

Originally built to the design of Wilson Worsdell for the North Eastern Railway, Class G5 (NER O) 0-4-2T no. 67322 was one of a trio sent to Cambridge by the LNER for employment on the Audley End to Bartlow service. The engine stands against the platform at the latter station on 25th August 1956, shortly before withdrawal. Photograph by R.C. Riley courtesy Rail-Online.

Above **BARTLOW STATION**
The Colchester, Stour Valley, Sudbury & Halstead Railway was authorised to build a line from Marks Tey (west of Colchester) to Clare and Bury St Edmunds in the early 1840s. Yet, the project stalled at Sudbury. Several amalgamations followed before the GER pushed the project forward to completion, in addition to extending to Cambridge. Bartlow station was on the latter section and was also the point for the diversion of the Saffron Walden branch which connected to the main line at Audley End. This quiet scene was captured at Bartlow station on 30th July 1961. Photograph by B.W.L. Brooksbank.

Below **BARTLOW STATION – NO. 61636**
Gresley B17 Class 4-6-0 no. 61636 *Harlaxton Manor* has a local train at Bartlow station in the 1950s. Constructed at Darlington in July 1931, no. 61636 worked through to the end of 1959. Photograph courtesy Rail-Online.

Above **BEDFORD MIDLAND ROAD STATION – NO. 75055**
The Railway Correspondence & Travel Society's London branch organised a trip through Essex, Cambridgeshire, Bedfordshire and Hertfordshire on 10th August 1958. Several locomotives were involved, including Hill J19 Class 0-6-0 no. 64656, Worsdell J15 Class 0-6-0 no. 65440, BR Standard Class 4 2-6-4T no. 80041, Johnson 3F Class 0-6-0 no. 43245 and BR Standard Class 4 4-6-0 no. 75055. The latter has the train at Bedford Midland Road station which was the point of pick-up for the locomotive, taking over from no. 80041. No. 75055 worked to Harpenden where the party left the main line for a jaunt on the branch to Hemel Hempstead, then returned for the final leg to St Pancras. Photograph by B.W.L. Brooksbank.

Opposite above **BEDFORD MIDLAND ROAD STATION**
On the formation of the Midland Railway in the mid-1840s, the company had an agreement with the London & Birmingham Railway to reach the capital. Problems soon arose and the decision was taken to enter a new arrangement with the Great Northern Railway, joining that company's line by building a connection between Leicester and Hitchin. Opened in 1857, Bedford did not receive a dedicated station on the line until 1859. Within three years of this date, the relationship with the GNR had deteriorated, leading to the construction of the new main line to St Pancras. Bedford station had the addition 'Midland Road' to the title following Grouping to distinguish from the ex-London & North Western Railway's station which also became part of the London Midland & Scottish Railway. This view of the Midland Road station frontage dates from 4th June 1962. Photograph by B.W.L. Brooksbank.

Opposite below **BEDFORD ST JOHNS STATION**
Some 13 years before the Midland Railway station opened, Bedford was connected to the main line to Rugby via a branch from Bletchley. Originally a terminus, the LNWR continued an earlier project to forge eastward to Cambridge and this opened to traffic from 1862. A cross-country route between Oxford and Cambridge was completed as a result and was subsequently known as the 'Varsity Line'. Today, the Bletchley-Bedford line is the only part operational and St Johns is again a terminus, though resited from the original location, as an unstaffed halt. Photograph by B.W.L. Brooksbank.

Above **BEDFORD ST JOHNS STATION – NO. 75055**
An express freight is behind BR Standard Class 4 no. 75055 at Bedford St Johns station, around 1960. The locomotive entered service from Swindon Works in January 1957 and was new to Bedford, joining five other class members there. In the main, they found use on commuter trains on the main line to London. Photograph courtesy Rail-Online.

Below **BELTON & BURGH STATION – NO. 62546**
Just to the south west of Yarmouth, Belton & Burgh station was open between 1859 and 1959 for the local communities at Belton with Browston and Burgh Castle. In July 1955, Holden D16/3 Class 4-4-0 no. 62546 *Claud Hamilton* has a local train for Ipswich. Photograph courtesy Rail-Online.

Above and below **BIGGLESWADE STATION**
On the East Coast Main Line, Biggleswade station was opened by the Great Northern Railway on 7th August 1850. Despite serving a relatively small population – around 4,000 – at the time, Biggleswade was upgraded at the turn of the century, boasting two island platforms, two signal boxes and a goods yard with shed, seen above to the right. When this image was captured in June 1956, the population had doubled and has since reached 15,000 people. The station continues to operate and was refurbished in the late 2000s, with around four trains stopping per hour at present. Both photographs by B.W.L. Brooksbank.

Below **BISHOP'S STORTFORD STATION – NO. 92010**

On 29th September 1955, BR Standard Class 9F 2-10-0 no. 92010 has reached Bishop's Stortford station with a March to Temple Mills freight train. The locomotive entered traffic from Crewe Works in May 1954 and new to March shed, remaining on the roster there to February 1957 when a transfer to Annesley occurred. This turned into the engine's longest allocation, spanning six years. Several moves took place subsequently to withdrawal in April 1966. Photograph by George C. Lander courtesy Rail Photoprints.

Above BISHOP'S STORTFORD STATION – NO. 70001

The Northern & Eastern Railway was originally proposed as a main line northward to reach York, yet the heavy financial burden of such an undertaking was too much and just two sections were completed. The first reached Broxbourne in 1840, then Stortford – later Bishop's Stortford – in 1842. The line was leased by the Eastern Counties Railway which forged northward to Cambridge and later connected with Norwich. In addition to being on this main route, Bishop's Stortford became the point of departure for a branch to Braintree, which after several delays, was completed in 1869. At the latter, a junction was made with the Braintree branch from the main line to Ipswich, Norwich, etc. BR Standard Class 7 'Britannia' Pacific no. 70001 *Lord Hurcomb* has an express – likely from Cambridge – at Bishop's Stortford on 13th July 1958. Photograph by K.C.H. Fairey courtesy Colour-Rail.

Above **BISHOP'S STORTFORD STATION – NO. 62551**
A local train has arrived with Holden D16/3 Class 4-4-0 no. 62551 at Bishop's Stortford on 16th August 1951. The locomotive was built at Stratford Works in November 1906 and worked until July 1956. At the time of the image, no. 62551 was allocated to Cambridge shed. Photograph by M.J. Reade courtesy Colour-Rail.

Below **BOTTISHAM & LODE STATION**
The first section of the line from Cambridge to Mildenhall included Bottisham & Lode station when opened in 1884. At this time the name was just Bottisham, with Lode added in 1897. The station closed in June 1962, with this picture taken a year later, though freight traffic remained active for another 12 months. Photograph by B.W.L. Brooksbank.

Above **BRAINTREE STATION – NO. 67189**
Holden F5 Class 2-4-2T no. 67189 has stopped at Braintree station with a local train in October 1955. Erected at Stratford in June 1903, the locomotive was in traffic to December 1956 and under BR was employed at Colchester depot. Photograph courtesy Rail-Online.

Below **BRAINTREE & BOCKING STATION**
The original station at Braintree opened with the completion of the branch from Witham on the main line. This was replaced by the present facility when the route from Bishop's Stortford was completed in the late 1860s. From 1910 until 1969 the station was officially referred to as Braintree & Bocking. Photograph by B.W.L. Brooksbank.

Above BRENTWOOD – NO. 61973
Travelling underneath the 1,500V DC overhead lines for the suburban electrification scheme from Liverpool Street to Shenfield is Gresley K3 Class 2-6-0 no. 61973. This project started under the LNER in the late 1930s and completed following the War in 1949. No. 61973 is near Shenfield at Brentwood with an express on 18th July 1952. At this time, the engine was based at Lowestoft and the allocation spanned over a decade, 1948-1959. Photograph by D. Preston courtesy Colour-Rail.

Opposite above BRAINTREE STATION – NO. 67228
T.W. Worsdell introduced the M15 Class 2-4-2T in the mid-1880s and his successor James Holden perpetuated the design to the early 20th century when 160 examples were in existence. Holden's son, Stephen Dewar Holden, was Locomotive Engineer of the GER between 1908 and 1912 and modified the design to create the G69 2-4-2T. Towards the end of his tenure 20 were built at Stratford Works, with no. 67228 completed in June 1911 as GER no. 69. The locomotive is in BR service here at Braintree station during the late 1950s. Stratford shed's '30A' code is on the smokebox door and this lasted much of the second half of the decade. Withdrawal occurred in April 1958. Photograph by P. Moffat courtesy Colour-Rail.

Opposite below BRAINTREE SHED – NO. 67214
A small straight shed with one line was erected just to the east of Braintree station in 1848. This was timber-built and survived until 1909 when damaged by strong winds. A new building was made with bricks by the GER soon afterwards and used until the late 1950s. Holden F4 Class 2-4-2T no. 67214 is outside the structure in the early 1950s, with London & North Eastern Railway branding still in place. The BR number was applied from May 1949. A J69 Class 0-6-0T stands behind inside the shed. The number is partially visible – no. 6862X. Photograph courtesy Rail Photoprints.

Below **BRUNDALL STATION – NO. 67216**
The Yarmouth & Norwich Railway was completed in time for the summer of 1844 and one of the five stations between the two places was Brundall. The line originally followed an indirect route but this was later changed by the GER which laid a line to Yarmouth via Acle making Brundall a junction. A train for Norwich has made a stop at Brundall during July 1955, with Holden F5 Class 2-4-2T no. 67216 piloting an unidentified D16/3 Class 4-4-0. No. 67216 was one of the class members fitted with armour plating for coastal defence trains during the Second World War. The locomotive was in service until the end of 1956 and had been employed at Lowestoft from the late 1940s. Photograph courtesy Rail-Online.

Above **BRENTWOOD – NO. 70006**

Eastbound from Liverpool Street station, BR Standard Class 7 'Britannia' Pacific no. 70006 *Robert Burns* has an express at Brentwood on 5th July 1959. The class was introduced as part of BR's push to modernise the locomotive stock across the system. A pressing need for suitable motive power was felt particularly across East Anglia as Thompson's B1 and Gresley's B17 Class 4-6-0s struggled with train weights. The 'Britannia' Pacifics appeared in time to take over the principal expresses in the area during the early 1950s and a number worked there to the early 1960s when dieselisation was completed. No. 70006 arrived at Stratford initially but was later sent to Norwich and remained there for ten years. In 1961, a transfer to March occurred, followed two years later by a move across country to Carlisle Kingmoor. Withdrawal from the latter took place in May 1967. The line to Brentwood offered the first challenge for eastbound trains, as from Harold Wood the ground rose for two miles at 1 in 103. Photograph by Hugh Ballantyne courtesy Rail Photoprints.

Opposite above BURY ST EDMUNDS STATION
– NO. 61631

The line from Ipswich to Bury St Edmunds was ready for traffic at the end of 1846, yet the western terminus was a temporary structure. Bury St Edmunds station was completed in late 1847 to the design of Sancton Wood, who also designed several others in the area, as well as further afield, such as Ireland. The buildings at Bury St Edmunds still stand and in the last ten years have undergone refurbishment. A mixed train is at the station during the mid-1950s with Gresley B17 Class 4-6-0 no. 61631 *Serlby Hall*. Photograph courtesy Rail-Online.

Opposite below BURWELL STATION

Around halfway on the Cambridge to Mildenhall line was Burwell station. This was open from 1884 to 1962 when the passenger service was withdrawn, though freight lingered to 1965. Burwell is pictured on 7th June 1963. The site has been cleared following closure. Photograph by B.W.L. Brooksbank.

Below BURY ST EDMUNDS SHED
– NO. 61555

The first locomotive shed at Bury St Edmunds was established in 1846 when the line from Ipswich opened. Around ten years later, the route extended westward and the shed was rebuilt further to the west of the station. This building survived until the early 20th century when the three-track shed seen here was erected and continued in use until 1960. Holden B12 Class 4-6-0 no. 61555 stands in the yard on 26th May 1957 – just five months remained for the engine before condemned. Photograph by K.C.H. Fairey courtesy Colour-Rail.

Above BURY ST EDMUNDS STATION – NO. 62613
At the start of the 20th century, a new express passenger locomotive was introduced by the GER. This was the S46 Class, later known as the 'Claud Hamilton', which was perpetuated throughout the first decade to reach over 100 examples. A final batch of ten locomotives was ordered at the cusp of Grouping and no. 62613 was amongst this number. A feature of these engines from the earlier class members was an enlarged boiler. No. 62613 is at Bury St Edmunds station with the 'Eastern Counties' railtour on 12th July 1959. Photograph courtesy Rail-Online.

Opposite above CAMBRIDGE STATION – NO. 61619
During 1953 Gresley B17 Class 4-6-0 no. 61619 *Welbeck Abbey* has a local train at Cambridge. Until a year previously, the engine had been employed at the city's locomotive depot and moved on to March shed, with the '31B' code on the smokebox door here. Photograph courtesy Rail Photoprints.

Opposite below BURY ST EDMUNDS STATION – NO. 67397
For the Great Northern Railway's local traffic in Yorkshire and London, H.A. Ivatt designed the C2 – LNER C12 – Class 4-4-2T. No. 67397, which was recently displaced from Hull, is at Bury St Edmunds station here with a train to Long Melford on 21st June 1955. Photograph by George C. Lander courtesy Rail Photoprints.

Below **CAMBRIDGE STATION – NO. 70038**
The Northern & Eastern Railway extended as far as Bishop's Stortford in 1842 before the project stalled. Taken over by the Eastern Counties Railway, an extension was built by that company through Cambridge to Brandon where a connection was made with the line to Norwich. Cambridge station, which was designed by Sancton Wood, subsequently became a local hub for a number of lines, including: Mildenhall branch; Bury St Edmunds branch; line to Huntingdon; LNWR line to Bedford; GNR line to Hitchin; Stour Valley route. BR 'Britannia' Pacific no. 70038 *Robin Hood* has a main line train for Liverpool Street passing through Cambridge station on 22nd June 1955. Photograph by George C. Lander courtesy Rail Photoprints.

Above CAMBRIDGE STATION – NO. 65460

Worsdell J15 Class 0-6-0 no. 65460 is working one of the last passenger trains on the Mildenhall branch. The final day was 15th June 1962 though freight lingered on for another two years. The line had been lightly used for a number of years previously and improvements in the form of diesel railbuses failed to overturn the decline. No. 65460 was similarly out of time as withdrawal from Stratford shed occurred in September 1962. The locomotive had served 50 years at this point. Photograph courtesy Colour-Rail.

CAMBRIDGE STATION – NO. 70039

An express is at Cambridge station with 'Britannia' Pacific no. 70039 *Sir Christopher Wren*. With '32A' on the smokebox door, the date of the image is between January 1959 and December 1960 when employed by Norwich shed. Photograph by Bill Reed.

Above **CAMBRIDGE SHED – NO. 61577**
Over the servicing pits at Cambridge shed is Holden B12 Class 4-6-0 no. 61577. The engine's five months at the depot was the last allocation of a career spanning 31 years. Photograph by Bill Reed.

Below **CAMBRIDGE – NO. 62035**
Many of the Peppercorn K1 2-6-0s used by the Eastern Region were based at March shed for freight traffic across East Anglia. Yet, the type was also welcomed by crews for employment on passenger trains and no. 62035 has a local here at Cambridge around 1960. Photograph by Bill Reed.

CAMBRIDGE STATION – NO. 61834
On 16th June 1960, Gresley K3 2-6-0 no. 61834 waits patiently for the tender to be filled at Cambridge station. Photograph by R.C. Riley courtesy Rail-Online.

Above **CAMBRIDGE SHED – NO. 64658**
Standing in the yard at Cambridge shed towards the end of the 1950s is Hill J19 Class 0-6-0 no. 64658. The locomotive did not see in the new decade as withdrawal occurred in November 1959. Photograph by Bill Reed.

Below **CAMBRIDGE SHED – NO. 64654**
The original shed at Cambridge was a short distance to the south of the station. Opened in 1845, a second depot was built in 1847 at the north end of the station and the pair existed together until 1865 when the first was closed by the GER. The company made substantial improvements to the building in 1913 and the LNER modernised in the early 1930s. Hill J19 Class 0-6-0 no. 64654 stands outside the structure during the late 1950s. Photograph by Bill Reed.

Above **CAMBRIDGE SHED – NO. 61236 AND NO. 63725**
Thompson B1 no. 61236 is on the left whilst Thompson O1 2-8-0 no. 63725 stands to the right as the pair are serviced in the yard at Cambridge shed on 2nd October 1960. The first mentioned worked from the depot and the latter was visiting from March. Photograph by B.W.L. Brooksbank.

Below **CAMBRIDGE SHED – NO. 62503**
Holden D15 Class 4-4-0 no. 62503 was one of the first D14 class members built in the early 20th century. At this time, the locomotive had a round-top firebox, though in the mid-1920s the Belpaire type was substituted, which also resulted in reclassification from LNER D14 to D15. The engine is at Cambridge shed on 29th August 1949. Withdrawal took place in 1951. Photograph by B.W.L. Brooksbank.

Above **CAMBRIDGE STATION – NO. 62527**
View from the northern end of the platform at Cambridge station as D16/3 Class no. 62527 departs with a local train on 25th August 1949. The engine was based at the local depot, which is visible in the background – Holden J67 0-6-0T no. 68587 stands in the yard on the right. Photograph by B.W.L. Brooksbank.

Below **CAMBRIDGE STATION – NO. 5356**
Worsdell J15 Class 0-6-0 no. 5356 still has the LNER number and branding here on 20th August 1949. The locomotive went on to receive BR no. 65356 in late 1950 and worked through to April 1957. No. 5356 has a branch train from Mildenhall at Cambridge station. Photograph by B.W.L. Brooksbank.

CAMBRIDGE SHED – NO. 61568
Norwich-based Holden B12 Class no. 61568 is at Cambridge shed for servicing in the late 1950s. The engine was condemned in August 1959. Photograph by Bill Reed.

Above CAMBRIDGE SHED – NO. E2794
The relatively rare 'E' prefix has been applied to Holden E4 Class 2-4-0 no. E2794. This was done just after Nationalisation before a new numbering system was devised by BR and denoted 'Eastern Region'. The locomotive is at Cambridge shed on 20th August 1949 and was renumbered in April 1950. Photograph by B.W.L. Brooksbank.

Below CAMBRIDGE SHED
The mechanical coaler at Cambridge shed was erected as part of modernisations carried out on the site by the LNER in the early 1930s. This image dates from 28th February 1951 and has been taken from the coaler looking down to the ash pits. Photograph by B.W.L. Brooksbank.

Above CHELMSFORD STATION – NO. 70006

The second section of the Eastern Counties Railway ran from Brentwood to Colchester and was ready for traffic from 1843. Just east of Brentwood, Chelmsford station opened at this time. The line approached the town (city after 2012) from the south before curving away eastward from the centre. Originally in a more northern position, the station was built on a viaduct which started as the railway crossed the River Cam. Before the end of the century, the GER improved facilities and moved the station to the present site. BR 'Britannia' Pacific no. 70006 *Robert Burns* passes through Chelmsford station with a Norwich to Liverpool Street express on 19th July 1951. The locomotive had been constructed at Crewe Works just three months earlier and was new to Stratford. A month later a move to Norwich occurred and the engine was there to 1961. Photograph by George C. Lander courtesy Rail Photoprints.

Below **CHELMSFORD – NO. 70038**
Around two miles north of Chelmsford, BR 'Britannia' Pacific no. 70038 *Robin Hood* has just passed New Hall signal box with the 11.33 express from Liverpool Street to Clacton. Visibly work-worn on 3rd October 1959, the engine had joined the ranks at Norwich earlier in the year following six years in employment at Stratford. Arriving there new in January 1953, no. 70038 was constructed at Crewe Works as part of the second order for 20 locomotives. When the last of these entered service some 23 class members were employed on the East of England lines, with 13 at Stratford and ten based in Norwich. In 1960, the situation had changed dramatically as Stratford had lost all 'Britannias' whilst another six were added to the total at Norwich. Photograph by K.L. Cook from Rail Archive Stephenson courtesy Rail-Online.

Below **COLCHESTER – NO. 62558**

The GER S46 Class 'Claud Hamilton' 4-4-0 reached 41 examples in the early 20th century. A major change occurred in 1903 when the firebox was changed to the Belpaire type and 70 of these D56 Class locomotives appeared from Stratford. As GER no. 1847, this engine was amongst these class members and was completed in December 1906. Twenty years later, as LNER no. 8847, the locomotive received a superheater which resulted in a change of LNER Class from D15 to D16/1. At this time the D16/1s had a short smokebox, though this was later lengthened and another sub-class created – D16/2. In 1933, a start was made on rebuilding D14 and D15 engines with a superheater and round-top firebox, with no. 8847, which had been renumbered 2558 in 1946, modified during September 1948. Around 1956, no. 62558 has a local train leaving Colchester. The engine was condemned in May 1957. Photograph from the John Day Collection courtesy Rail Photoprints.

Above COLCHESTER STATION – NO. 70039
In May 1958, 'Britannia' Pacific no. 70039 *Sir Christopher Wren* has made a stop at Colchester station. The class was introduced in 1951 when the country was celebrating the Festival of Britain. As a result, a naming policy was selected to honour prominent people from history, though there were exceptions. For example, 'Britannias' that went to the Western Region took the names from withdrawn 'Star' Class locomotives. An accident in that area also resulted in modification to the front end. The derailment of no. 70026 *Polar Star* at Milton killed 11 and injured 157 and a contributing factor was the handrails on the smoke deflector obscuring the view forward. The rails were replaced by hand holes which are installed here above and to the right of the nameplate. No. 70039 was Stratford-allocated but early in 1959 transferred to Norwich. Photograph courtesy Colour-Rail.

Above COLCHESTER – NO. 61648

In the mid-1920s, improvements in coaching stock for boat trains resulted in the B12 4-6-0s and D16 4-4-0s struggling. H.N. Gresley was hard-pressed to design a suitable locomotive that provided the power whilst meeting stringent bridge and permanent way requirements. The task was subsequently left to the North British Locomotive Company, yet the new locomotives – the B17 Class 4-6-0 – were still off the specifications and a request for further orders was denied by the LNER. Seventy-one were ultimately completed, with some built to work on the ex-Great Central Railway main line. No. 61648 *Arsenal* was built at Darlington in March 1936 and reported for work at Leicester. In September 1939, the locomotive, as LNER no. 2848, moved over to the Great Eastern Section and remained in the area to withdrawal in December 1958. The engine is at Colchester here in May 1958. Picture by F. Hornby courtesy Colour-Rail.

Below **COLCHESTER STATION – NO. 65443**
The opening of the Great Northern & Great Eastern Joint line left the latter company with an urgent requirement for suitable freight locomotives. T.W. Worsdell introduced the GER Y14 Class 0-6-0 in 1883 and around 60 were built before he left the railway for the North Eastern. His successor James Holden added to this number, as well as S.D. Holden and A.J. Hill up to 1913 when 289 had been put into traffic. No. 65443 was built at Stratford before the end of the 19th century and worked to the end of 1959. The locomotive is light engine at Colchester station a month before condemned. Photograph by A. Morris courtesy Colour-Rail.

Above **COLCHESTER STATION – NO. 69700**
Colchester was the terminus at the end of the second phase of the Eastern Counties Railway when opened in 1843. Despite the plan to reach Ipswich and Norwich, the company took over the Northern & Eastern Railway in the following year and decided to extend to the Norfolk Railway which gave access to Norwich. The people of Ipswich were understandably upset by this development and formed the Eastern Union Railway to connect Colchester and Ipswich. This line opened in 1846. Colchester became a junction soon after as independent companies were formed to reach Hythe, Walton-on-the-Naze and later Clacton-on-Sea. The main line was north of Colchester and when the extensions to the south west were made, a second station was built – St Botolph's, later Colchester Town. Hill N7 Class 0-6-2T no. 69700 has paused at Colchester main line station with a local train for Clacton-on-Sea during the 1950s. The engine had an allocation to Stratford spanning the decade and withdrawal occurred in December 1960. Photograph courtesy Rail-Online.

Opposite **COLCHESTER STATION – NO. 61667**
One result of the restrictions imposed on the design of the B17 Class was the coupling of a particularly small tender. This allowed the engines to use the 50 ft turntables that dominated the Great Eastern Area. The tenders had stepped-out tops, 3,700-gallon water capacity and could carry four tons of coal. When later class members were destined for the Great Central Area, LNER group standard tenders with straight sides were provided, having a 4,200-gallon water space and 7½-ton coal capacity. No. 61667 *Bradford* has the latter type here, being new to Woodford Halse in April 1937. Yet, this was not the original as a switch had occurred with another group standard tender in the early 1950s. A local train is coupled to no. 61667 at Colchester station during May 1958. Photograph by F. Hornby courtesy Colour-Rail.

COLCHESTER – NO. 61670

B17 no. 61670 *City of London* has the 11.20 Yarmouth South Town to Liverpool Street station near Colchester on 5th September 1959. Photograph by K.L. Cook from Rail Archive Stephenson courtesy Rail-Online.

Above **CRESSING STATION – NO. 65470**
In the mid-1840s, a short line from Maldon on the Essex coast to Braintree was authorised and completed in 1848. Bulford station opened just to the south of the last mentioned, though was later renamed Cressing in 1911. A local train is pictured there in the 1950s with J15 no. 65470. The signal box has since been preserved at the Colne Valley Railway. Photograph courtesy Rail-Online.

Below **CROMER BEACH STATION – NO. 43145**
Ivatt Class 4MT 2-6-0 no. 43145 is with a local train to Melton Constable at Cromer Beach station in July 1956. Photograph courtesy Rail-Online.

Above **CROMER BEACH STATION – NO. 61113**
Ian Allan founded his publishing company in the mid-1940s following the success of the 'ABC Guide' of the Southern Railway's locomotives. The first magazine from Ian Allan appeared in 1946 and this was *Trains Illustrated* which ran to 1962 when retitled *Modern Railways*. A number of railtours were organised through the magazine, as well as a dedicated club, and one is at Cromer Beach station on 19th September 1954. This had originated at Liverpool Street with 'Britannia' Pacific no. 70000 and travelled on the main line to Norwich where Thompson B1 no. 61113 took over to Cromer, then Melton Constable. Holden B12 Class 4-6-0 no. 61530 was placed at the head of the special there and journeyed via Spalding and Sleaford to Grantham then returning to London on the East Coast Main Line. Photograph by J.B. McCann courtesy Colour-Rail.

Opposite above **CROMER BEACH STATION – NO. 70035**
The East Norfolk Railway built the first line to Cromer, reaching the town gradually from Norwich by 1877. This was a short distance away from the settlement, though the Midland & Great Northern Joint Railway later built a station in the town – Cromer Beach. A connection was later made between the two lines. BR Standard Class 7 Pacific no. 70035 *Rudyard Kipling* is at the station in July 1955. Services to the original station – Cromer High – had ceased during the previous year. Photograph courtesy Rail-Online.

Opposite below **ELSTREE – NO. 42587**
The 15.40 local train from St Pancras to St Albans is in the Elstree area on 8th August 1959. Stanier 4P Class 2-6-4T no. 42587 is the locomotive employed and had recently transferred to St Albans from London, just four months earlier. The engine was only there until early 1960, as no. 42587 migrated northward to Derby. Photograph by M.J. Reade courtesy Colour-Rail.

Above ELY STATION – NO. 62510
No. 62510 was built at Stratford in July 1900 as the last engine from the first order for S46 Class 4-4-0s. As GER no. 1899, the locomotive was the only one from this first batch to be rebuilt with a saturated boiler, whilst the remainder had a superheated boiler fitted. Under the LNER, the engine was superheated in 1933 and transformed to D16/3 specifications during 1943, but retained the original slide valves. No. 62510 has a train at Ely station during the 1950s. Photograph courtesy Rail-Online.

Opposite above ELSTREE – NO. 44235
The Kentish Town breakdown crane has been called out on 3rd October 1959 and is seen at Elstree. Being the main passenger depot in London for the Midland Main Line, Kentish Town was a strategic position for one of several breakdown cranes ordered for the line in the late 1930s. These had 50 tons capacity and replaced several from earlier in the decade which had 35 tons capacity, though some were upgraded from the latter at the time. Fowler 4F Class 0-6-0 no. 44235 is in charge of the crane at Elstree. Arriving at Kentish Town earlier in 1959 from the Somerset & Dorset Joint Line, the locomotive worked in London to August 1962 when transferred to Wellingborough. Photograph by M.J. Reade courtesy Colour-Rail.

Opposite below ELY STATION – NO. 70030
Whilst many of the early 'Britannia' Class Pacifics were allocated to the East of England services, the sphere of operations soon expanded to include the Southern, Western and London Midland Regions. No. 70030 *William Wordsworth* went to Holyhead for the North Wales coast trains, yet the water capacity of the tender was too low for the duties and a move to Longsight, Manchester, occurred a month after entering traffic in November 1952. There, 'Britannias' were used on services between Manchester and Euston. This allocation lasted to mid-1953 when no. 70030 joined the ranks at Norwich. The engine has the 09.09 Liverpool Street to Yarmouth Vauxhall express passing through Ely station on 16th August 1958. Photograph by B.W.L. Brooksbank.

Above **ELY – NO. 62522**
A southbound local train is near Ely on 26th April 1958 with D16/3 Class no. 62522. The locomotive was built in March 1902 and rebuilt twice under the LNER in 1924 and 1938. Working from King's Lynn when pictured, the engine transferred to Cambridge shortly after, though was condemned in August. Photograph by R.C. Riley courtesy Rail-Online.

Opposite above **ELY STATION – NO. 62558**
Edward Thompson instigated the reboilering of the D14 and D15 Class 4-4-0s when he was Mechanical Engineer, Stratford, during the early 1930s. With his initial experiment deemed a success, the programme expanded to providing new cylinder blocks with piston valves. Yet, this addition proved detrimental to the frames and frequent instances of cracks could not be rectified. When the rebuilding resumed after the war, no further conversions to piston valves took place and the valances over the driving wheels were also retained whereas removal had occurred pre-1939. No. 62558 started life as D56 (LNER D15) no. 1847 and was converted to D16/1 in 1926. The engine was amongst six to change to D16/3 in 1948, whilst the final six were modified in the following year. No. 62558 retained both slide valves and valancing. The locomotive has a train from Ely to King's Lynn in the 1950s here. Photograph courtesy Rail-Online.

Opposite below **ELY STATION – NO. 65438**
Following the Eastern Counties Railway's takeover of the Northern & Eastern Railway, the line from Cambridge to Brandon was built with Ely located around halfway on the route. Opened in 1845, the station was designed by Sancton Wood and Francis Thompson and cost a considerable amount – approx. £80,000. By the end of the decade, a line from King's Lynn had been constructed and joined the Norwich route just north of Ely, whilst the ECR expanded westward via March to Peterborough. Branches to St Ives and Newmarket were later connected to Ely. The LNER upgraded the station facilities after Grouping and further work was carried out in the 1990s. During the late 1950s, Worsdell J15 Class no. 65438 has a train arrived at Ely from Fordham, which was a short distance north of Newmarket. Photograph courtesy Rail-Online.

Above FEN DITTON – NO. 62796
A total of 100 Holden T26 Class 2-4-0s were constructed between 1891-1902 for GER mixed traffic duties. Of this number, only 18 were in traffic for the formation of BR in 1948. No. 62796 was one of these engines and has a local train to Mildenhall at Fen Ditton on 13th October 1956. The engine survived until May 1957, whilst the final T26 – LNER E4 – was withdrawn in 1959. Photograph courtesy Rail-Online.

Opposite above ELY STATION – NO. 43089
The Hunstanton portion of the 10.39 ex-Liverpool Street express stands by a platform at Ely station on 16th August 1958. The locomotive is Ivatt Class 4MT 2-6-0 no. 43089 which was one of several dozen class members drafted to East England during the early 1950s, mainly to take over from the ageing Midland & Great Northern Joint stock. Originally a London Midland & Scottish Railway design, BR soon recognised the versatility of the type and redeployed the 4MT into new areas. No. 43089 was new to Peterborough in December 1950, though moved over to Neasden six months later. In September 1954, the locomotive transferred to Cambridge and when pictured had reached King's Lynn. Photograph by B.W.L. Brooksbank.

Opposite below FAKENHAM WEST STATION – NO. 61545
As a competitor to the Great Eastern Railway, the Lynn & Fakenham Railway was promoted during the late 1870s. This ran from King's Lynn to Melton Constable where a split was made, going southwards to Norwich and eastwards to North Walsham. At the latter, a connection was made with the Yarmouth & North Norfolk Light Railway to access Yarmouth. Fakenham station was opened on the line from August 1880, being known as Fakenham Town to differentiate from the GER station on the route to Wells. The Lynn & Fakenham Railway was one of several companies to form the Eastern & Midlands Railway in 1881 which was later part of the Midland & Great Northern Joint Railway. BR renamed both Fakenham stations after Nationalisation, with Fakenham Town becoming Fakenham West and this remained in use to closure in 1959. Some three years before this date, in July 1956, B12 Class 4-6-0 no. 61545 has Yarmouth to Birmingham service paused at the station. Photograph courtesy Rail-Online.

Below **FORDHAM STATION – NO. 64777**
One of the largest classes built by the LNER was Gresley's J39 Class 0-6-0. Numbering 289 examples, the duties were mainly freight related, though some passenger work took place over the years. The J39s appeared between 1926 and 1941, with around 180 employed in the Eastern Area. On the ex-GER lines, the class mainly operated from Stratford, Ipswich, Norwich and March. Twenty ordered for the region before the 'unification of brakes' programme in the late 1920s had Westinghouse brakes from new and unlike similar engines in the North East, this equipment was retained throughout. No. 64777, as LNER no. 2724, was one of these locomotives when built at Darlington Works in April 1929. The first allocation was Ipswich, but soon moved on to Parkeston. Remaining in the GE area to withdrawal in January 1960, no. 64777 is passing through Fordham station on 27th July 1954 with the 13.20 Whitemoor to Parkeston freight service. Photograph courtesy Rail-Online.

Above **FORDHAM STATION – NO. 65460**
Amongst several stations on the Ely & Newmarket Railway was Fordham which opened as Fordham and Burwell on 1st September 1879. Five years later, the station became a junction to the new line from Cambridge to Mildenhall and a change to just Fordham occurred at the same time. In June 1962, Worsdell J15 Class no. 65460 is seen again working one of the final passenger services on the Mildenhall branch, having paused briefly at Fordham. The latter remained open for another three years, although the line continues to operate as the Ely-Ipswich route. Photograph courtesy Colour-Rail.

Above **HAUGHLEY STATION – NO. 61580**

For three years, Haughley Road station was just a stop on the line from Ipswich to Bury St Edmunds. This line was backed by the Eastern Union Railway and following opening in 1846, the company soon sought an extension to Norwich. Diverging northward at Haughley, the station was rebuilt as a result, also being renamed Haughley Junction from July 1849. Later, Haughley Road, Haughley and Haughley West were used before Haughley was settled on as the name from the early 1930s. An Ipswich-bound local train is leaving Haughley station on 4th May 1958. The locomotive is Holden B12 Class 4-6-0 no. 61580, which was the last of ten built for the LNER in 1928, also being the final class member. Photograph by R.C. Riley courtesy Rail-Online.

Opposite above **HAUGHLEY STATION – NO. 62608**

Haughley station was rebuilt again under the GER during the 1860s and these facilities remained in use until 1967. In the early 20th century, the Mid-Suffolk Light Railway terminated on the east side of the station, running to Laxfield. Holden D16/3 Class no. 62608 has a local train, viewed from the footbridge in the mid-1950s. Upgraded from D15 to D16/3 in 1937, no. 62608 ran in this form to withdrawal from Cambridge in January 1957. Photograph courtesy Rail-Online.

Opposite below **HAUGHLEY STATION – NO. 64841**

Gresley J39 Class 0-6-0 no. 64841 was one of 25 engines ordered at the end of 1929. Appearing during June 1932, the locomotive was new to Glasgow Eastfield shed and remained there to 1943. The next allocations took LNER no. 2980, later 4841, to Aberdeen and Dundee before a swap occurred. This involved several Scottish engines with steam and vacuum brakes which were exchanged with vacuum-braked J39s from the GE Area. Initially concentrated at Carlisle in 1946, no. 64841 transferred to Ipswich in April 1947 and remained until condemned for scrap towards the end of 1959. The locomotive is passing through Haughley station with a freight train off the Bury St Edmunds line bound for Ipswich in the 1950s. Photograph courtesy Rail-Online.

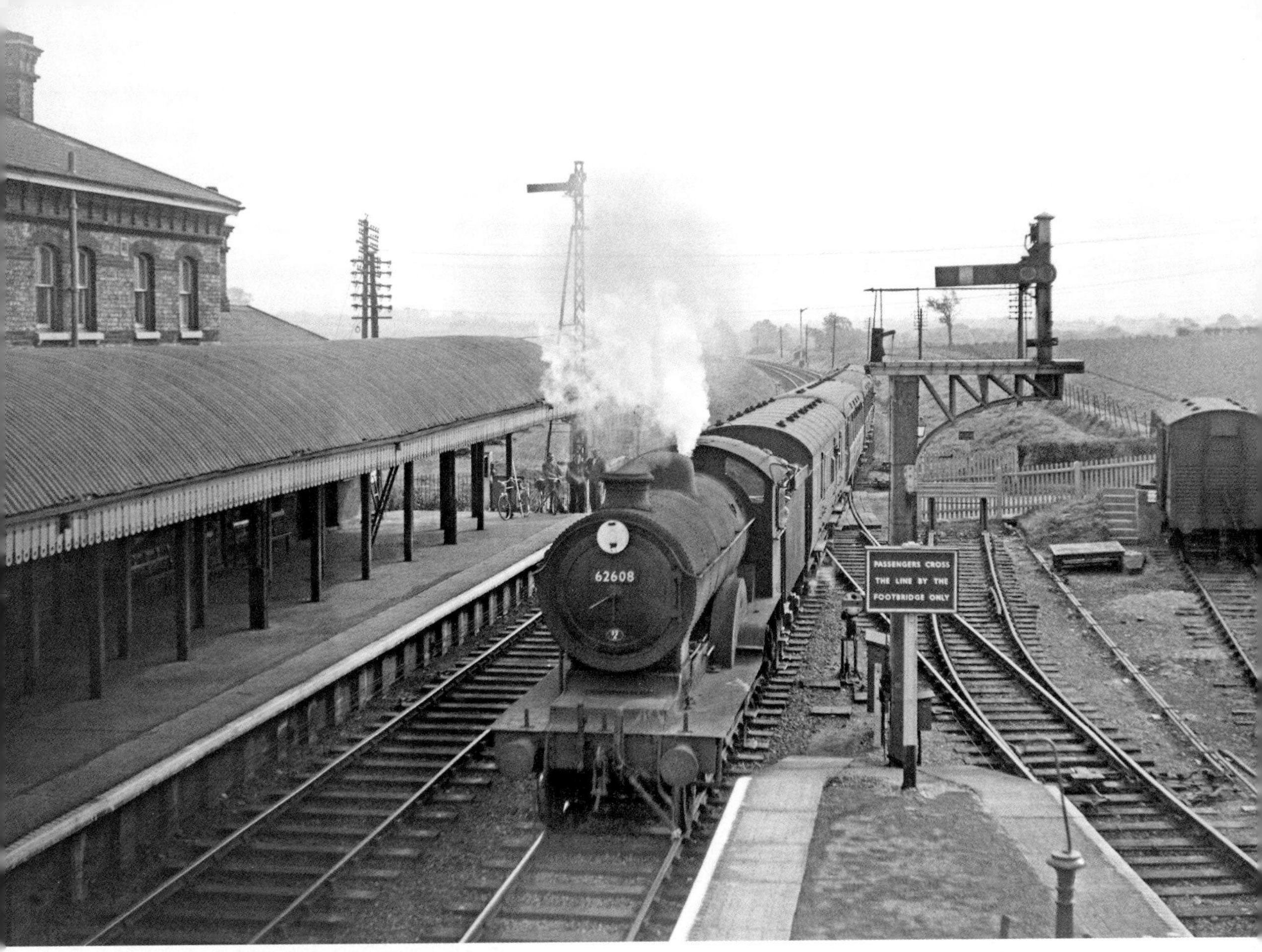
62608
PASSENGERS CROSS
THE LINE BY THE
FOOTBRIDGE ONLY

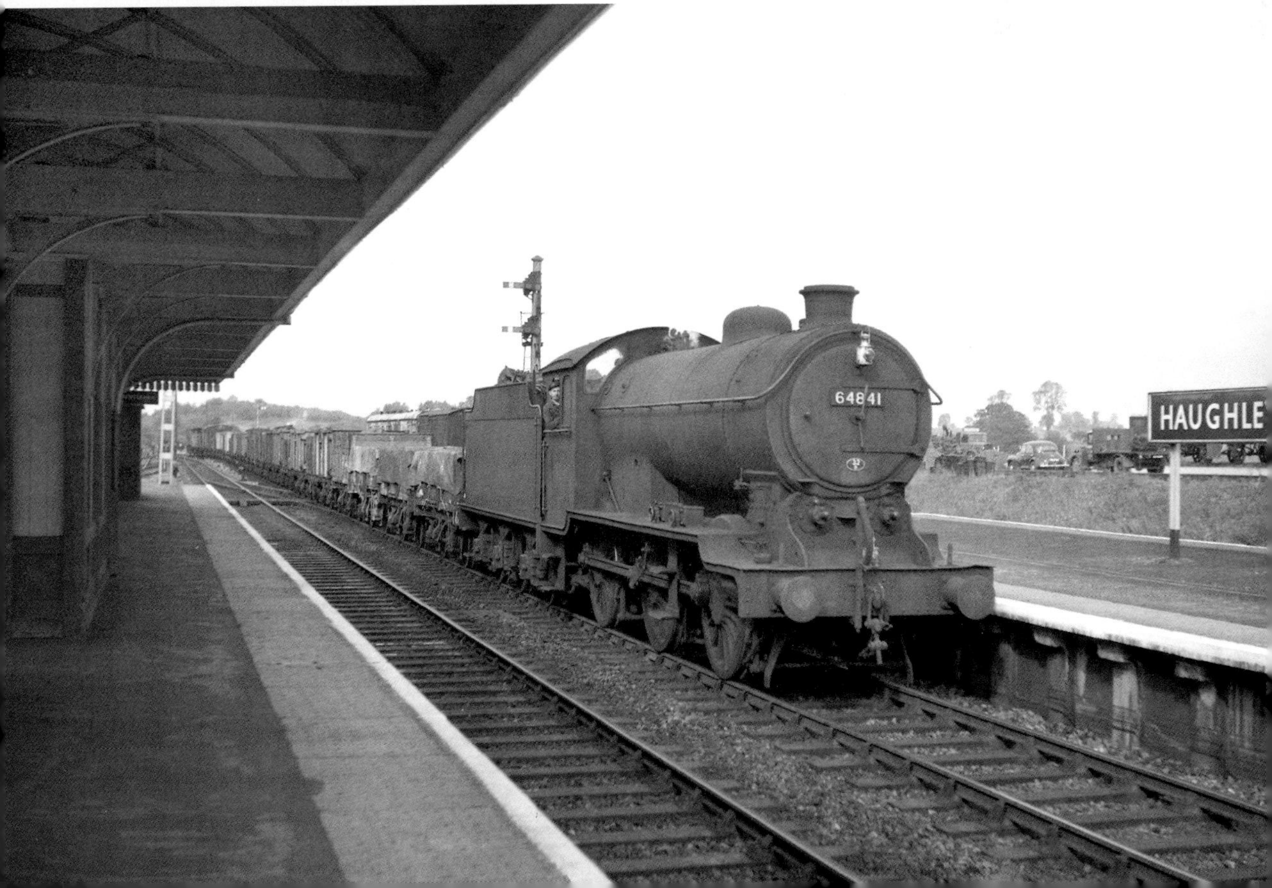
64841
HAUGHLE

Above HAVERHILL STATION – NO. 62795

The Holden E4 Class was particularly useful to the GER for around 30 years before the formation of the LNER when the type was deemed life-expired and relegated to secondary duties. In the mid-1930s, a number of E4s were able to prove their continuing capabilities by taking over for a time on the Darlington to Penrith 'Stainmore Line'. No. 62795, as LNER no. 7411, was dispatched to Kirkby Stephen in October 1935 for trains to Darlington. Working over the difficult terrain of this area, the weather was uncomfortable for the enginemen and in mid-1936 no. 7411 was fitted with a cab at Doncaster Works, whilst other class members on the line were similarly treated during the year. In January 1941 no. 7411 returned to the GE Area and later mainly worked at Cambridge to withdrawal in March 1955. This event was around a year away when no. 62795 was caught with this local service to Cambridge at Haverhill. Photograph from the John Day Collection courtesy Rail Photoprints.

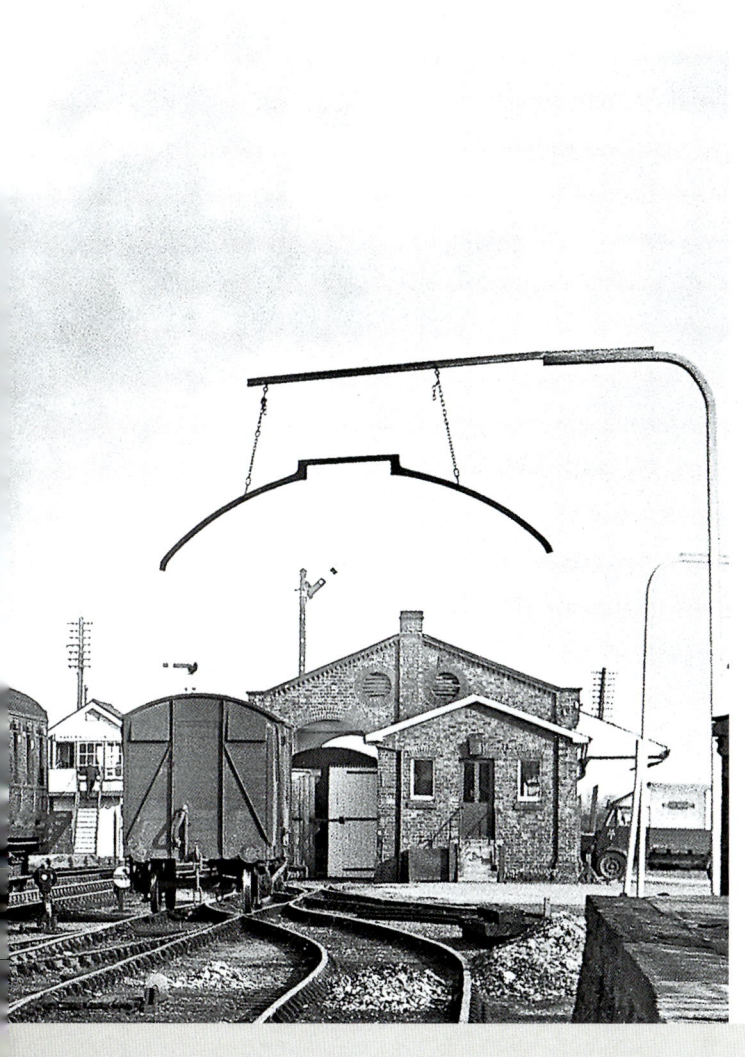

Below **HAVERHILL STATION – NO. 65478**

Founded in 1911, the Cambridge University Railway Club is one of oldest dedicated to the railways still in existence, whilst also enjoying longevity amongst societies at the university. The club has a railtour, which has gone unrecorded, at Haverhill station during April 1961. The locomotive is Cambridge-based Worsdell J15 Class 0-6-0 no. 65478. Photograph by Dave Cobbe courtesy Rail Photoprints.

Above **HAVERHILL STATION – NO. 62785**
Following the withdrawal of no. 62789 in late 1957, no. 62785 became the sole representative of the Holden E4 Class for the next two years. Over this period, the engine was employed at Cambridge for local stopping freight trains. The final duty for no. 62785 was organised by the Cambridge University Railway Club on 27th November 1959. No. 62785 was subsequently preserved for the nation and is presently at Bressingham. An earlier event has been captured on 27th April 1958 at Haverhill station, with the society engaging the locomotive for a railtour. Photograph by R.C. Riley courtesy Rail-Online.

Opposite above **HEMSBY STATION – NO. 62524**
Just north of Yarmouth, the village of Hemsby was provided with a station in 1878 as the Great Yarmouth & Stalham Light Railway forged northward, ultimately reaching North Walsham. The original station at Hemsby was a temporary structure as the company failed to obtain permission for a level crossing and this took two years to rectify, at which time a permanent station was built. This later became part of the Midland & Great Northern Joint Railway. D16/3 Class no. 62524 is running on the original section of line with the 18.07 from North Walsham to Yarmouth at Hemsby on 28th August 1958. Hemsby station closed on 2nd March 1959. Photograph by Hugh Ballantyne courtesy Rail Photoprints.

Opposite below **HAVERHILL STATION – NO. 65469**
The Marks Tey to Sudbury line was open for around 15 years before the extensions to Cambridge were finalised. The first section from Shelford to Haverhill was complete for 1st June 1865, then the connection to Sudbury was ready in August. The route then operated as the Stour Valley Railway to partial closure in 1967. Before the main route was constructed, the Colne Valley & Halstead Railway was built and eventually made a connection at Haverhill. A separate station serving that route resulted in name changes to Haverhill North and Haverhill South after Grouping, though the latter was closed to passengers in 1924. Haverhill North remained so named to 1952 and reverted to Haverhill to closure. Another Cambridge University Railway Club special is pictured here, with J15 no. 65469 at the head during May 1962. Photograph courtesy Colour-Rail.

Above HITCHIN STATION – NO. 90730
R.A. Riddles adapted Stanier's 8F Class 2-8-0 design for use by the War Department in the early 1940s. At the end of the conflict 935 had been erected for use at home and abroad. Amongst the final examples was no. 90730 when built at Vulcan Foundry in May 1945. At the end of the decade, the locomotive joined the ranks at Peterborough New England shed and was employed there to September 1962. During October 1957, no. 90730 has reached Hitchin with a loaded coal train. Photograph courtesy Colour-Rail.

Opposite above HILLINGTON STATION – NO. 43154
To the east of King's Lynn, Hillington station opened on the line to Fakenham in August 1879. Doncaster-built Ivatt Class 4MT no. 43154 has the 09.30 Peterborough to Yarmouth train there on 30th August 1958. Working from Melton Constable at this time, in March 1959 a transfer to Colwick occurred. Withdrawal from there took place in December 1964. Photograph by Hugh Ballantyne courtesy Rail Photoprints.

Opposite below HITCHIN STATION – NO. 43766
Samuel Waite Johnson oversaw the locomotive department of the Midland Railway for 30 years. Of the many types introduced over this period, a high proportion were 0-6-0 goods locomotives. A particularly large number were built in the last ten years of Johnson's tenure, with no. 43766 amongst those appearing just before his retirement. The engine was part of the 2736 series which numbered ten initially constructed at Derby Works in 1903 – later, another 40 were produced. No. 43766 was in traffic from January 1903 and for much of the engine's later career, Bedford had the allocation. The locomotive is at Hitchin station during June 1960. Photograph courtesy Colour-Rail.

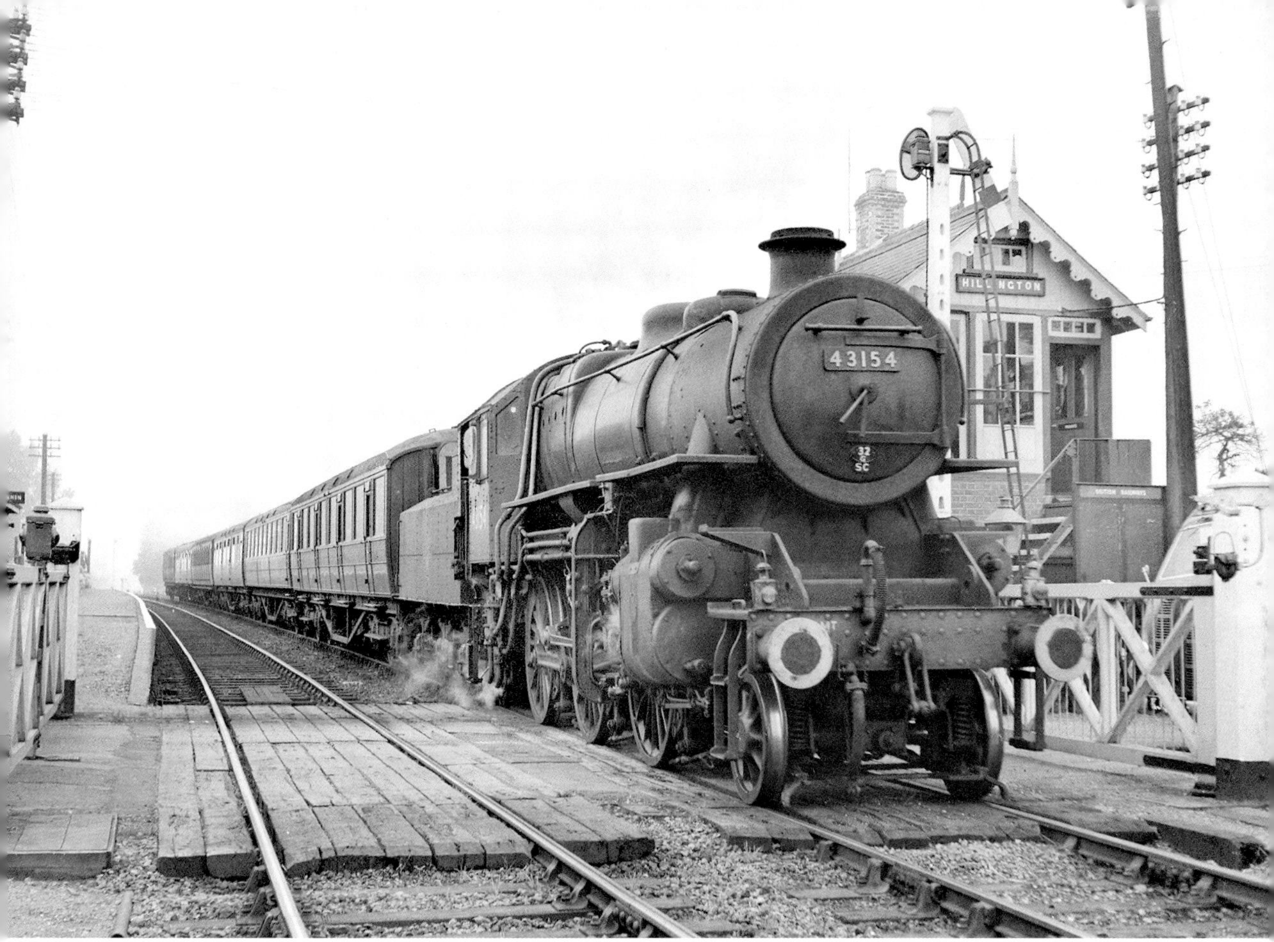
HILLINGTON
43154
32
SC

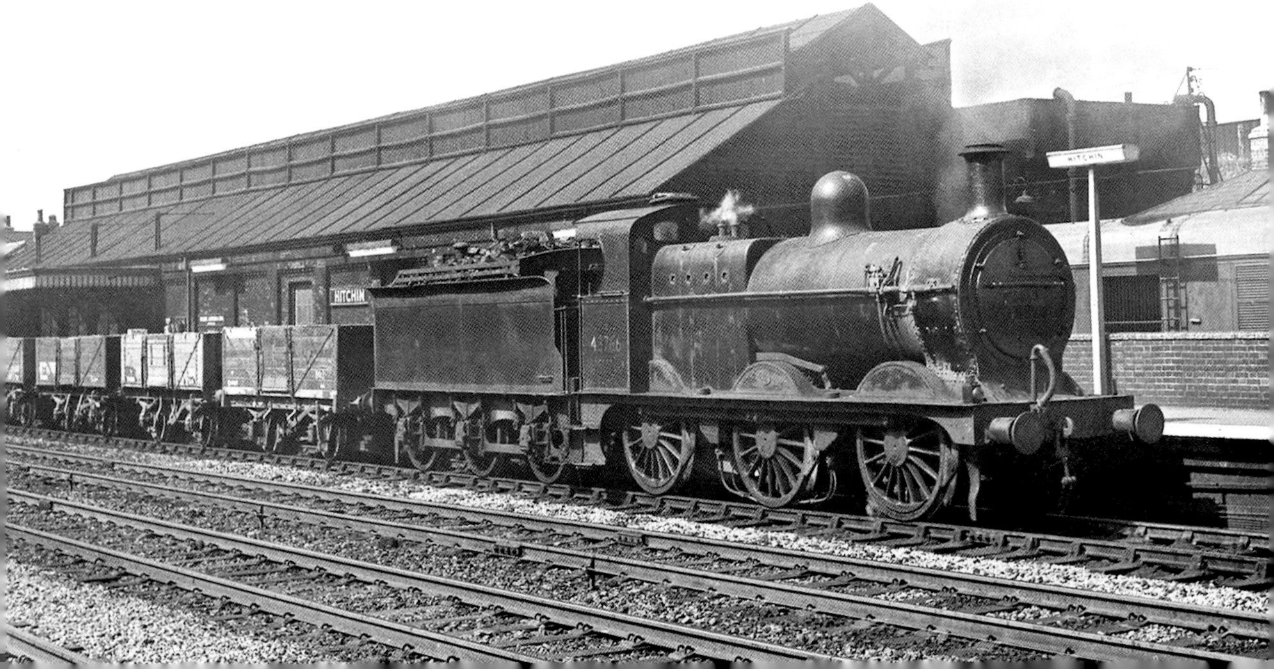
HITCHIN
HITCHIN

Above HITCHIN STATION – NO. 41270
The rise of the motor car in the post-war period irreparably damaged local rail traffic. Various schemes were attempted to save money on the services, such as railbuses introduced in the late 1950s. The Hitchin-Bedford line had these trialled at the time, yet passenger trains were withdrawn in late 1961. Ivatt Class 2MT no. 41270 has a local train for Bedford at Hitchin station on 30th April 1955. New to Bedford in 1950, the locomotive was there for eight years. Photograph by B.W.L. Brooksbank.

Opposite above HITCHIN STATION – NO. 60506
Thompson A2/2 Class Pacific no. 60506 *Wolf of Badenoch* started life as Gresley P2 Class 2-8-2 no. 2006. The engine was one of six erected for use on the line between Edinburgh and Aberdeen. Towards the end of the Second World War, Edward Thompson rebuilt all as Pacifics and the class subsequently relocated to England. No. 60506 arrived at Peterborough in 1949 and remained there until condemned during April 1961. The locomotive is at the head of a local train in April 1956. Photograph by F. Hornby courtesy Colour-Rail.

Opposite below HITCHIN STATION – NO. 43808 AND NO. 43766
As mentioned, the line from Bedford to Hitchin initially offered the Midland Railway a route to London before the company's own route was built. When the switchover occurred, traffic declined markedly and did not recover to closure in the early 1960s. Services were mainly freight transfer, with a brick train arriving here on 2nd November 1957. The wagons are headed by a pair of ex-MR 0-6-0s, with the lead engine being Deeley 3F no. 43808 and Johnson 3F no. 43766 behind. Photograph by B.W.L. Brooksbank.

HUNTINGDON STATION – NO. 67744
The 18.00 Huntingdon to Hitchin local is ready to depart from the first mentioned station on 23rd July 1959. The locomotive is Thompson L1 Class 2-6-4T no. 67744.
Photograph by D.C. Ovenden courtesy Colour-Rail.

Above HUNTINGDON STATION – NO. 60055

Fifty-eight miles from London, Gresley A3 Class Pacific no. 60055 *Woolwinder* starts a southbound service from Huntingdon station in 1957. Photograph by G.B. Wallis courtesy Rail Photoprints.

Below HUNTINGDON STATION – NO. 92179

A long train of coal wagons trail behind BR Standard Class 9F 2-10-0 no. 92179 which travels through Huntingdon station on 22nd July 1958. The locomotive was new to Peterborough around a year earlier and the allocation lasted eight years. Photograph by D.C. Ovenden courtesy Colour-Rail.

IPSWICH STATION – NO. 62552
A local train arrives at Ipswich station behind D16/3 Class no. 62552 in 1952. Photograph by John Chalcraft courtesy Rail Photoprints.

Above **IPSWICH STATION – NO. 62783**
Passengers have alighted from this train originating at Bury St Edmunds and E4 Class 2-4-0 no. 62783 is reversing out of Ipswich station to run around the set before taking up the next duty. Pictured on 16th September 1952, the engine was in traffic for another two years. Photograph by George C. Lander courtesy Rail Photoprints.

Below **IPSWICH STATION – NO. 61645**
The Saturday 11.01 service from Gorleston-on-Sea to Liverpool Street speeds through Ipswich station with Gresley B17 Class no. 61645 *The Suffolk Regiment* on 7th July 1951. Photograph by B.W.L. Brooksbank.

Below IPSWICH STATION – NO. 70002
Immediately south east of Ipswich station, Stoke Hill provided an obstacle for the Eastern Union Railway joining with Colchester. The line's engineer Peter Schuyler Bruff drove a tunnel through the mound some 361 yards long which was distinguished for being the first on a particularly sharp curve. The earthworks also uncovered several interesting archaeological finds. BR 'Britannia' Pacific no. 70002 *Geoffrey Chaucer* is about to plunge into Stoke tunnel with this Norwich to Liverpool Street express on 16th September 1952. Photograph by George C. Lander courtesy Rail Photoprints.

Above ISLEHAM STATION – NO. 65438

Between Fordham and Mildenhall on the branch from Cambridge was Isleham station. According to disused-stations.org.uk, the main building stood on the eastbound platform, which was 360 ft long. The Cambridge platform was slightly longer by 20 ft and had a waiting room provided. J15 Class no. 65438 is stood at this platform during the mid-1950s. The engine worked from Cambridge between November 1952 and July 1958 when condemned. Isleham station was closed to goods in 1964 and the main building became a private residence, whilst part of the goods yard was taken over by a tyre business. Later, part of the track bed accommodated a swimming pool for the house, but the whole section has been filled and landscaped subsequently. Photograph courtesy Rail-Online.

Above **KING'S LYNN STATION – NO. 62614**
D16/3 Class no. 62614 is at King's Lynn station in the early 1950s. The engine was based locally from the mid-1940s to withdrawal during August 1958. Photograph courtesy Rail Photoprints.

Below **KING'S LYNN SHED – NO. 65501 AND NO. 68545**
On 20th May 1956, two locomotives have been captured in the yard at King's Lynn shed. On the left is Holden J69 Class 0-6-0T no. 68545 and to the right stands J16 Class 0-6-0 no. 65501. Photograph by B.W.L. Brooksbank.

Above **KING'S LYNN SHED – NO. 67386**
The first shed at King's Lynn was opened on a site to the east of the station in 1846 by the Lynn & Ely Railway. In use for around 25 years, the GER moved to another location further east where a four-road building was opened and occupied to 1959. On the left is Ivatt C12 Class 4-4-2T no. 67386 which was sent from Ardsley, Leeds, to King's Lynn in 1951 to work the connecting train between there and South Lynn. This duty was undertaken, along with a spell on the Yarmouth-Lowestoft line, to withdrawal in April 1958. To the right is D16/3 Class no. 62614 which is particularly well presented in comparison, owing to the fact the engine was kept as the Royal Engine to work services to Sandringham. Photograph courtesy Rail-Online.

Right KING'S LYNN STATION – NO. 64641
An excursion train leaves King's Lynn station in the 1950s behind Holden J18 Class 0-6-0 no. 64641. Photograph courtesy Rail-Online.

Opposite below KING'S LYNN STATION – NO. 67374
Initially built for West Riding traffic by the GNR, Ivatt's C12 Class 4-4-2T also found employment in London subsequently before being displaced to more areas. No. 67374 is another class member sent to South Lynn for the short connection to King's Lynn and has a push-pull set here in the mid-1950s. Photograph courtesy Rail-Online.

Below LINTON STATION – NO. 65391
As with many other large classes, the GER Y14 (LNER J15) specifications did not remain static over the construction period. The major change for these 0-6-0s occurred after seven years when the boiler pressure was raised by 20 lb per sq. in. to 160 lb per sq. in. and the grate was redesigned which allowed a slightly larger area to be used. No. 65391 was amongst the last class members with the first specification built in October 1890. Under the LNER, the two types were distinguished by different diagram numbers, 31 and 32 respectively. By Grouping, the diagram 31 was declared life-expired and spares were discontinued, resulting in the adoption of the diagram 32 for the 1883-1890 engines. No. 65391 had the boiler type changed in 1943. The locomotive has a Cambridge to Colchester train at Linton station – between Pampisford and Bartlow on the Stour Valley line – during October 1955. Photograph courtesy Rail-Online.

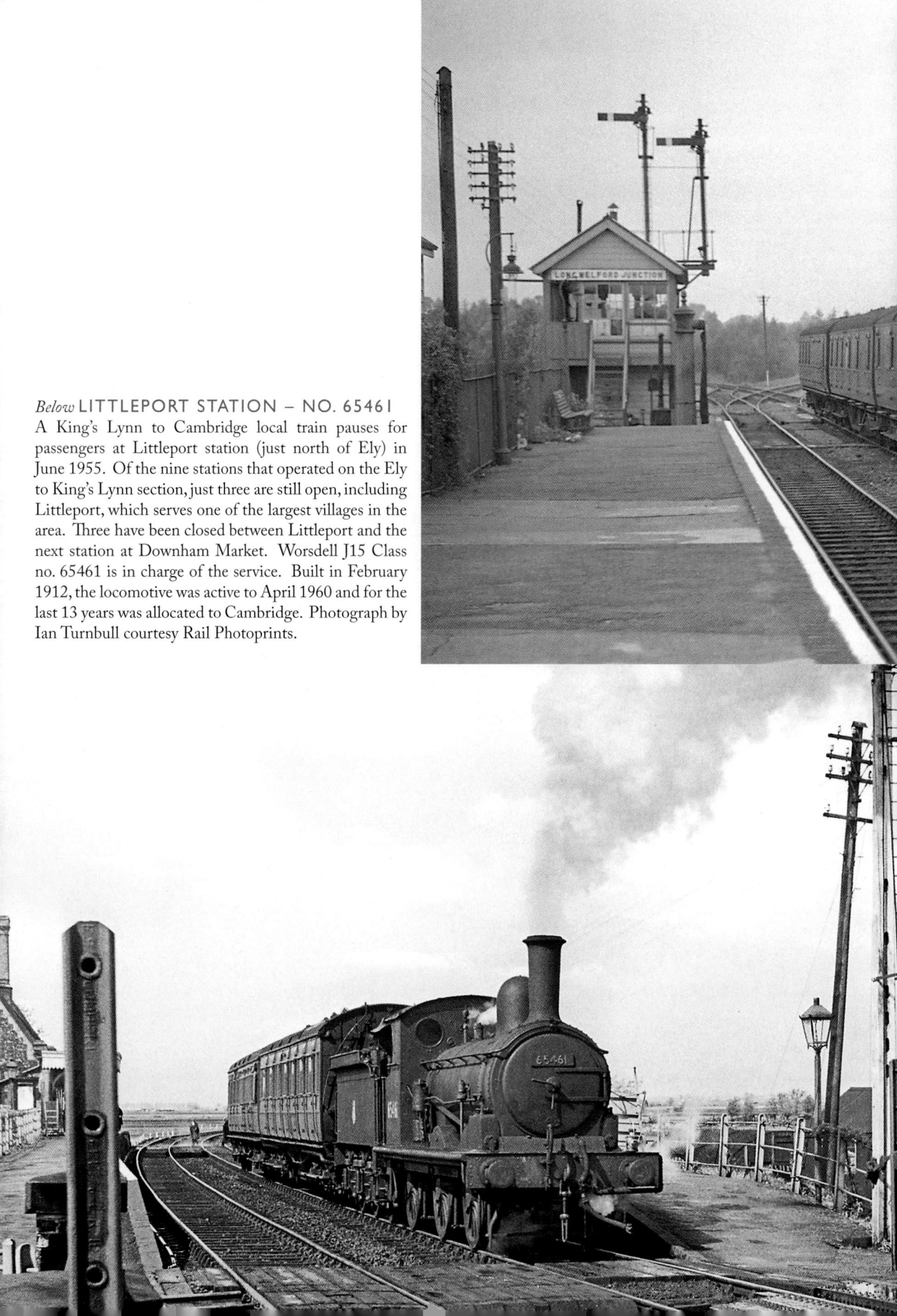

Below LITTLEPORT STATION – NO. 65461
A King's Lynn to Cambridge local train pauses for passengers at Littleport station (just north of Ely) in June 1955. Of the nine stations that operated on the Ely to King's Lynn section, just three are still open, including Littleport, which serves one of the largest villages in the area. Three have been closed between Littleport and the next station at Downham Market. Worsdell J15 Class no. 65461 is in charge of the service. Built in February 1912, the locomotive was active to April 1960 and for the last 13 years was allocated to Cambridge. Photograph by Ian Turnbull courtesy Rail Photoprints.

Above LONG MELFORD STATION – NO. 62786

Just to the north of Long Melford station, the line to Cambridge diverged from that to Bury St Edmunds. Long Melford was opened with the extension westward in 1865 as Melford, with the change in name occurring nearly 20 years later. When closed in the 1960s, the station buildings became a private residence, while the track bed was filled and new housing has been erected on parts of the site. Arriving at Long Melford with a Cambridge to Colchester train in August 1955 is E4 Class no. 62786. Constructed at Stratford Works in January 1895, the engine's career lasted over 60 years to July 1956. Photograph courtesy Rail-Online.

Below LOWESTOFT – NO. 37

The Sentinel Waggon Works began producing steam lorries in the early 20th century before diversifying into other areas, such as small steam locomotives and railcars. At Grouping, the LNER tested one of the company's four-wheel locomotives with a view to employing the type on light railways and in goods yards or places with operating restrictions. The Sentinel locomotive differed from the standard by using a chain drive from a vertical steam engine to the wheels rather than normal motion. Proving satisfactory in the trials, the first was ordered in 1925 and received into traffic by September. Y1 Class 0-4-0T no. 8400 was sent to Lowestoft for use on the docks and in the LNER sleeper yard, remaining employed there to January 1956. The locomotive carried a running number to May 1953 when taken into departmental stock as no. 37. This picture was taken during the mid-1950s in Commercial Road, Lowestoft, with the locomotive apparently shunting wagons between the docks. Photograph courtesy Rail-Online.

Above **LOWESTOFT CENTRAL STATION – NO. 67204**
The Somerleyton estate, to the north west of Lowestoft, was purchased by Samuel Morton Peto in the mid-1840s. The son of a builder, he embraced the new railways and was involved in the construction of several early projects. Whilst developing the estate, he became involved in the Yarmouth & Norwich Railway and soon after turned to connecting Lowestoft to the route at Reedham and improving the port facilities there. Opened in 1847, Lowestoft was subsequently transformed to cater for seafaring and the associated industries. This development saw the badly-sited original terminus rebuilt in the mid-1850s with an appropriate structure. Before the end of the century, Lowestoft welcomed increasing numbers of holidaymakers and shortly after the start of the 20th century, renaming to Lowestoft Central was carried out. The Norfolk & Suffolk Joint Railway had opened at this time between Yarmouth and Lowestoft resulting the change, with their facility called Lowestoft North. Shunting carriages at Lowestoft Central station on 18th July 1955 is Holden F4 Class 2-4-2T no. 67204. Based locally from May 1953, the locomotive's career was to end imminently during September. Interestingly, 'British Railways' is still displayed on the tank sides at this late date. Photograph courtesy Rail-Online.

Above **LOWESTOFT SLEEPER DEPOT – NO. 40**
The price of Britain ruling the high seas and feeding the workers of the industrial revolution was the severe depletion of the nation's forests. Trees were felled to create the ships of the Royal Navy before the switch to iron and steel vessels, whilst landowners changed the use of woodland to agricultural areas where money could be made from food production. As a result, around only 10% of the nation's timber requirements were met by home production, with the remainder fulfilled by imports. The situation reached such low levels during the First World War, the government was obliged to establish the Forestry Commission which set out to address the imbalance. The railways had to import wood for carriage construction (such as teak from Asia), as well as sleepers for the permanent way. The LNER operated several depots for sleepers across the country, mainly at coastal areas to streamline the production process. Lowestoft Sleeper Depot was established in the early 20th century to the south west of the station on the banks of the inner harbour. As mentioned, the first Sentinel locomotive was put to work there shortly after Grouping, whilst several others found employment there subsequently. Y3 Class no. 40 arrived after Nationalisation as no. 8173. Renumbered departmental 40 in 1953, the locomotive was at Lowestoft to withdrawal in May 1964. Several years earlier, on 22nd May 1957, the engine is amidst the large stacks of sleepers ready for dispatch. Photograph by R.C. Riley courtesy Rail-Online.

Opposite **LOWESTOFT SHED – NO. 68565**
Whilst the Lowestoft line was still being promoted, the Yarmouth & Norwich Railway joined forces with the Norwich & Brandon Railway to form the Norfolk Railway. The new company soon leased the Lowestoft line and oversaw construction. This included an engine shed just to the west, on the north side of the line, which was in use during the year following opening. Used for around 35 years, a new depot was constructed under the GER in the early 1880s. This was further west and on the south side of the tracks to Lowestoft station. The four-road building features here behind Holden J69 Class 0-6-0T no. 68565 on 22nd May 1957. The locomotive was at the shed for three years from March 1956 to November 1959. At this time around 20-30 engines were allocated to Lowestoft, being mainly freight locomotives. The depot was closed in mid-1962. Photograph by K.L. Cook from Rail Archive Stephenson courtesy Rail-Online.

Below **MANNINGTREE STATION – NO. 65453**

Between Colchester and Ipswich on the Eastern Union Railway's line, Manningtree station was opened by the company during 1846. Ten years later, an important junction was made for the line to Harwich, east of the station. At the turn of the century, Manningtree was rebuilt. A westbound local train looks to be leaving the station here on 30th August 1956. Worsdell J15 Class no. 65453 is in charge. Built in June 1906, the locomotive was in traffic to August 1962. From 1951, no. 65453 was based a short distance away at Parkeston. In January 1961 the engine relocated to Stratford. Holden J17 Class 0-6-0 no. 65513 is partly visible on the left. Photograph courtesy Rail-Online.

Above MARCH STATION – NO. 65539

A busy scene at March station from 16th August 1958. Holden J17 Class 0-6-0 no. 65539 has an eastbound freight, whilst a Gresley B17 Class 4-6-0 is at the platform in the background. Holden based his J16 (GER F48) Class on the 'Claud Hamilton' 4-4-0 which made the 0-6-0s one of the most powerful of the type in the country at the time. These initially had a round-top firebox, yet as an experiment one was fitted with a Belpaire firebox in 1902. No. 65539 was the engine and proved the suitability of the design. Following two more batches of J16 locomotives, the Belpaire firebox was adopted for the final 30 locomotives (GER Class G58) built from 1905 to 1911. At Grouping these were classified J17, with some earlier engines joining the class. Both J16 and J17 were superheated under the GER and LNER, with no. 65539 altered in October 1929. Photograph by B.W.L. Brooksbank.

Above MARCH STATION – NO. 44932
Stanier Class 5 no. 44932 stands with a parcels train at March station on 16th August 1958. Allocated to Agecroft shed, Manchester, at this date, the locomotive had perhaps been 'borrowed' or pressed into service by another shed on a short-term loan. Earlier in the day, the engine had been caught by the same photographer arriving with the 11.10 Birmingham to Clacton-on-Sea service. No. 44932 later relocated closer to March, being employed at Leicester and Annesley on the ex-Great Central Railway main line. Surviving to the end of steam in August 1968, no. 44932 was preserved and has just returned to the main line following an overhaul. Photograph by B.W.L. Brooksbank.

Opposite above MARCH SHED – NO. 63747
The Railway Operating Division of the British Army adopted J.G. Robinson's O4 (GCR 8K) Class 2-8-0 design during the war to help the movement of military materials at home and abroad. No. 63747 was amongst over 500 built for the ROD, emerging from the North British Locomotive Company's Queen's Park Works, Glasgow, in November 1917. The LNER subsequently purchased three batches, with no. 63747 part of the final order for 100 placed in 1927. Working mainly in Lincolnshire for the LNER, after Nationalisation the engine returned to the county following a spell at Mexborough. No. 63747 was Frodingham-allocated from then to withdrawal in May 1961 and is pictured around that time in the yard at March shed. Photograph by Bill Reed.

Opposite middle MARCH SHED – NO. 63924
Retford-based Gresley O2 Class 2-8-0 no. 63924 rests at March shed in the late 1950s. The engine was withdrawn in November 1963 with over 40 years in traffic. Photograph by Bill Reed.

Opposite below MARCH SHED – NO. 60803
No. 60803 was built as part of the first batch of Gresley's V2 2-6-2 Class in the mid-1930s. The engine was erected at Doncaster which produced just 25 of the 184 class members. No. 60803 is at March in the late 1950s. Photograph by Bill Reed.

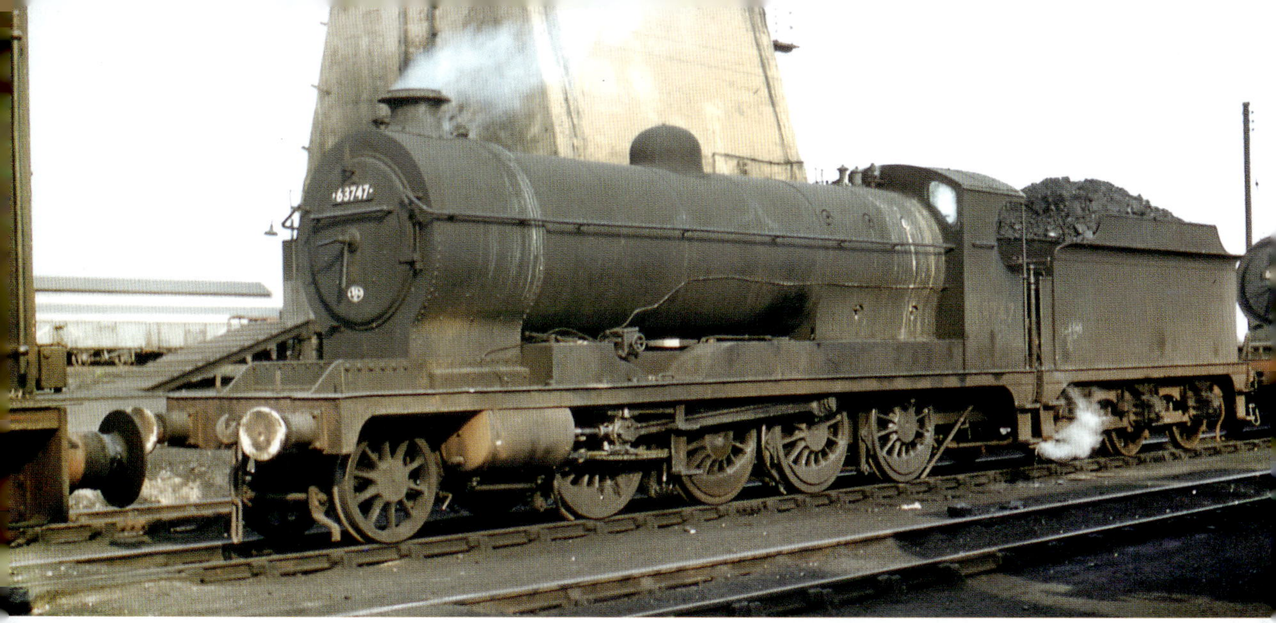
63747

63924

60803
60803

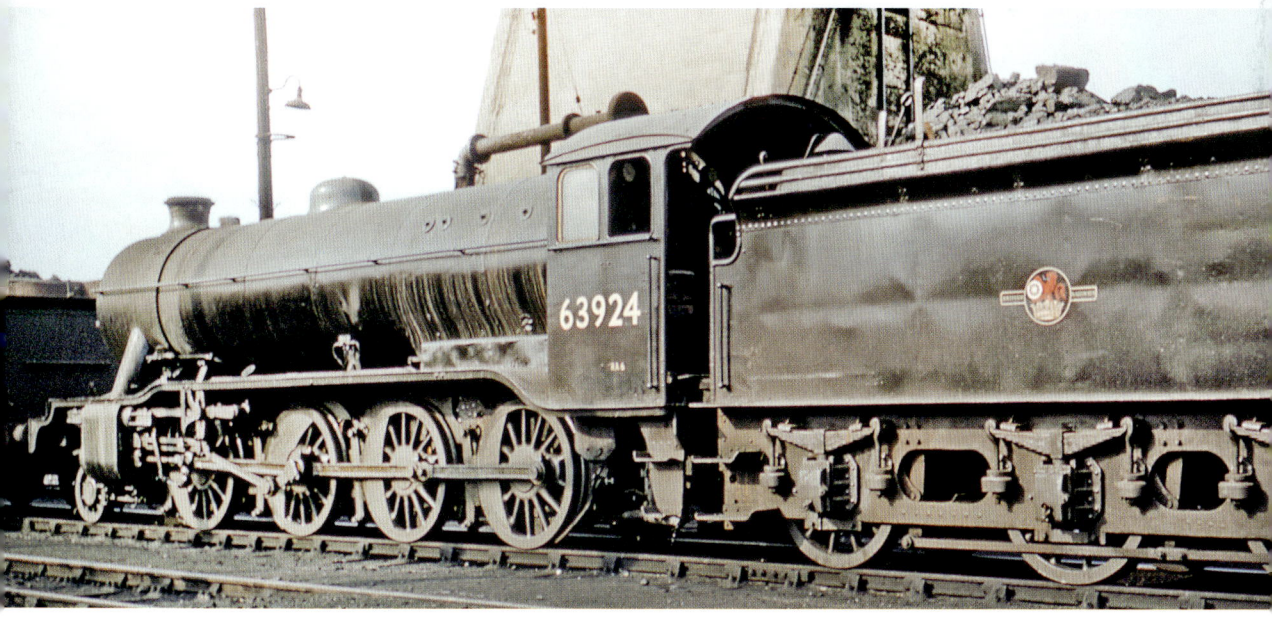

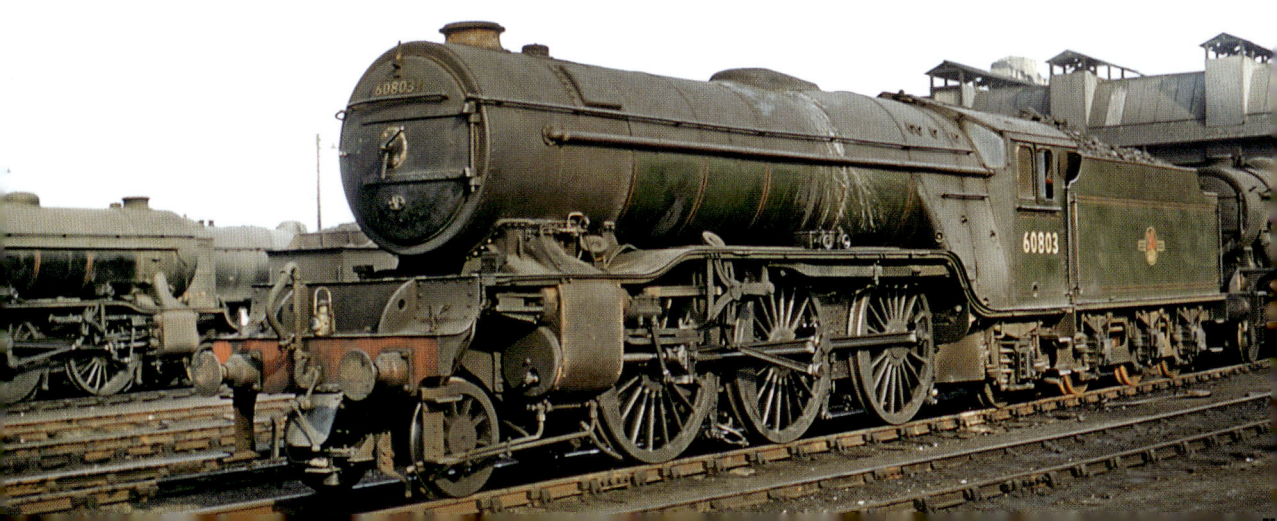

Above MARCH – NO. 64691

Like Holden before him, A.J. Hill adapted a passenger design for heavy freight duties. He took the B12 Class 4-6-0 and produced the J20 0-6-0, which took the power title from the earlier J16/J17 Classes. Just 25 of the J20s were constructed in the early 1920s and all saw rebuilding with a round-top firebox. No. 64691 emerged from Stratford in December 1922 and later changed from the Belpaire type in December 1944. In 1954, the locomotive joined the ranks at March and has been tasked with being the station pilot here on 16th August 1958. No. 64691 survived for another four years before sent for scrap. Photograph by B.W.L. Brooksbank.

Opposite above MARCH STATION – NO. 62530

March station was originally on the Eastern Counties Railway line from Ely to Peterborough and in use from 1847. During the following year, the station became a junction for the route to Wisbech, which later connected to the King's Lynn line. March became an important junction with the opening of the GN&GE Joint line in the 1880s and a large marshalling yard was built nearby. D16/3 Class no. 62530 has a Peterborough East to Cambridge stopping service leaving on 16th August 1958. Photograph by B.W.L. Brooksbank.

Opposite below MARCH – NO. 61642

View east from the B1101 level crossing near March station as Gresley B17 Class no. 61642 *Kilverstone Hall* approaches with the 14.26 Ely to Birmingham New Street on 16th August 1958. New to the GC Section from Darlington in 1933, the locomotive found employment at Gorton, with a brief stint at Neasden, to April 1937, then transferring to Cambridge. No. 61642 remained there to September 1958 when condemned. As the engine was originally allocated to the GC Section, a Group Standard tender was paired, though a short time after Nationalisation a GE-type tender was attached and coupled to the end. Photograph by B.W.L. Brooksbank.

Above MARCH – NO. 61890

One of Gresley's first designs was the K1 (GNR H3) Class 2-6-0 which appeared in 1912. Used on freight trains, these duties, particularly in the First World War, became taxing and Gresley had to increase the power reserves. These K3 (GNR H4) Class 2-6-0s had a 6 ft diameter boiler, which was the largest in use in Britain at the time, as well as three cylinders operated by Gresley's new valve gear. Ten were built initially before Grouping, then the design was chosen to be a Group Standard type for use across the system. A total of 193 were built up to 1937, with no. 61890 amongst nine produced at Darlington in 1930. The engine was the first of the series, which had some slight detail differences in the cab from other class members. No. 61890 was based at March shed from October 1953 to September 1962. The locomotive is approaching March station with a train of apparently empty wagons. Photograph by D. Collins courtesy Colour-Rail.

Below MARKS TEY STATION – NO. 61651

The 14.01 Colchester to Cambridge service pauses at Marks Tey station on 23rd August 1958. The station was the last on the Eastern Counties Railway main line to Colchester when completed in 1843. At the end of the decade, Marks Tey became a junction for the Sudbury branch and in the mid-1860s the extension to Cambridge. Gresley B17 Class no. 61651 *Derby County* has the local train here and was allocated to Colchester from December 1953 to February 1959. The engine then transferred to Cambridge though did not survive in service beyond the end of the year. Photograph by L. Rowe courtesy Colour-Rail.

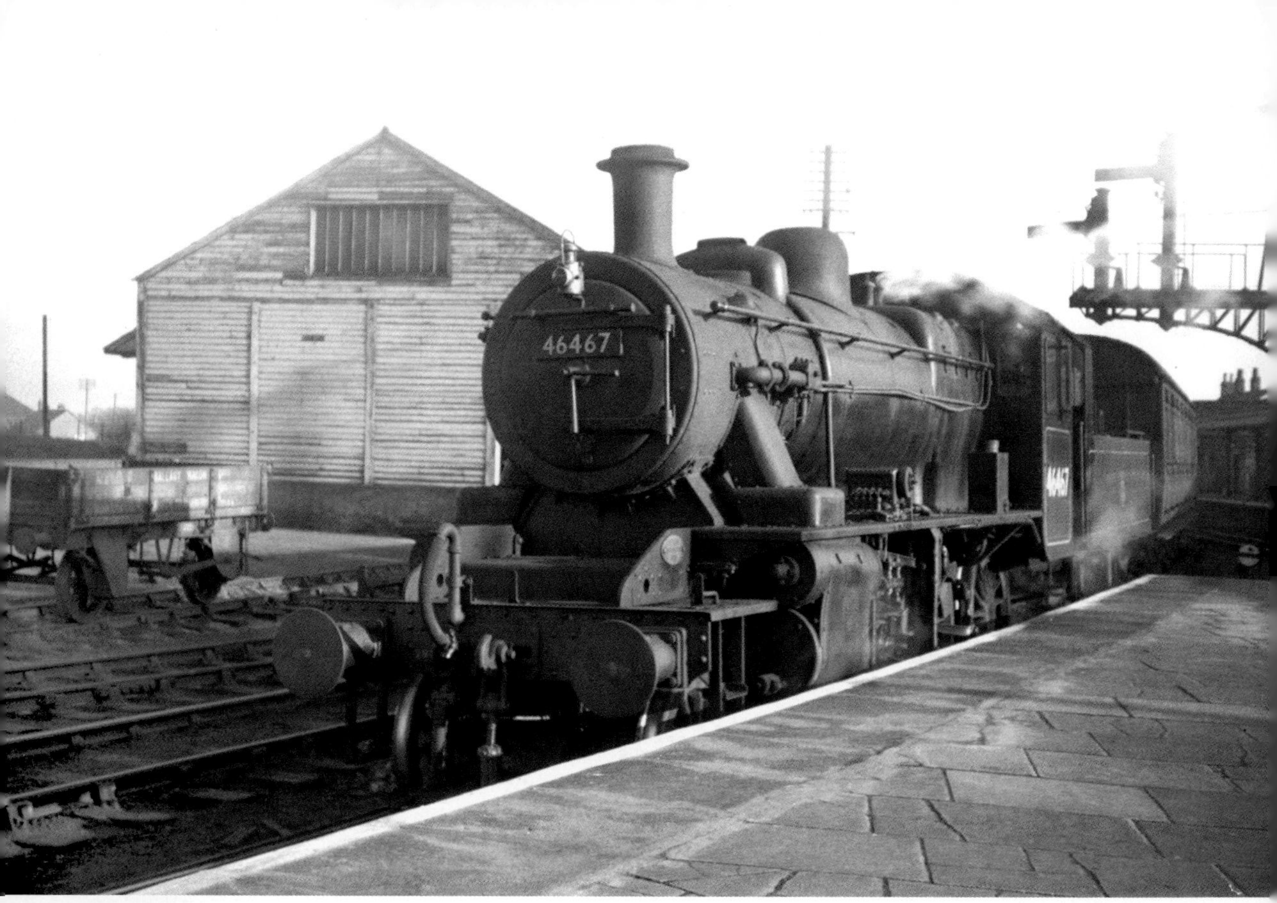
46467
46467

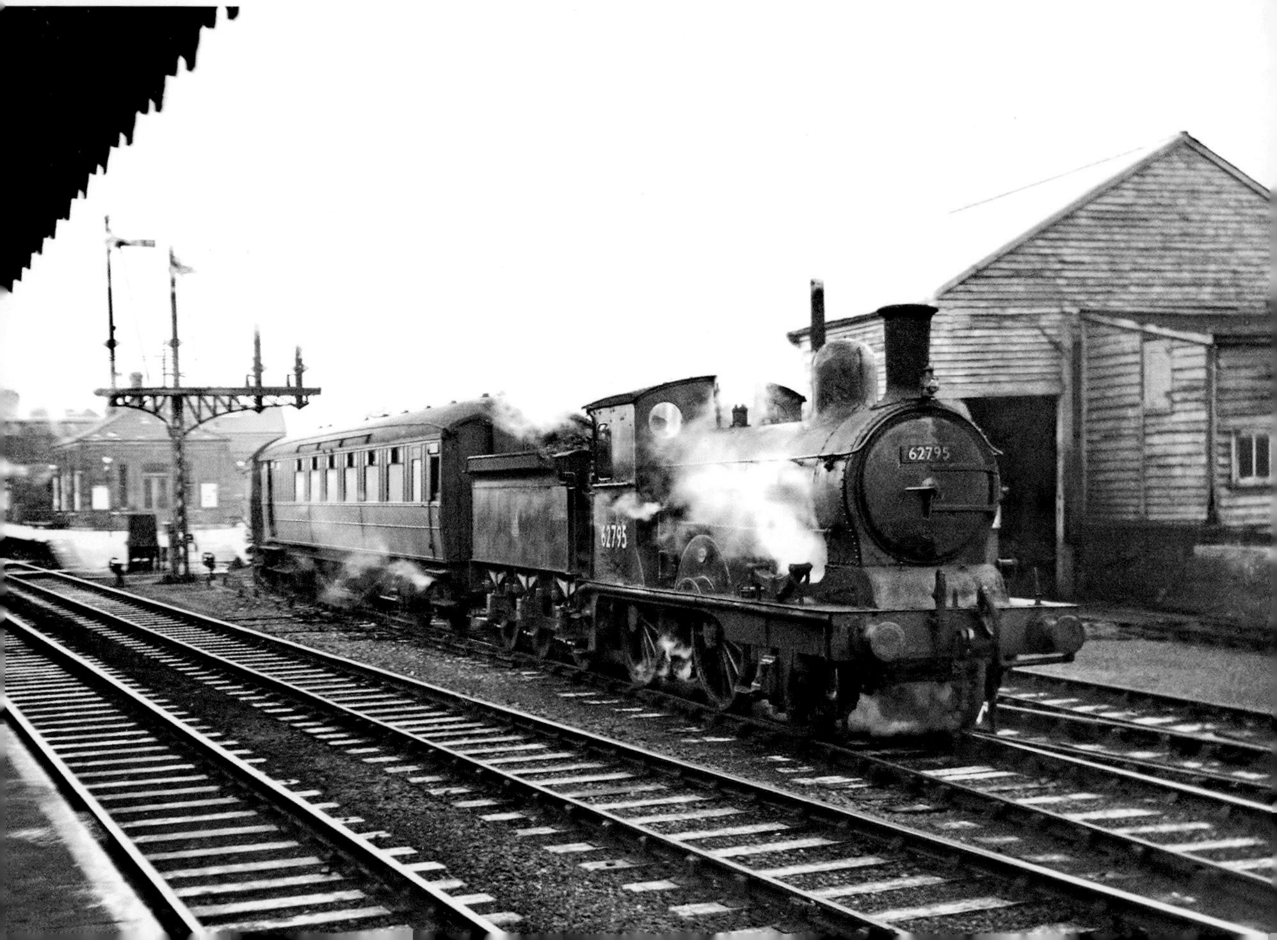
62795
62795

Above **MELTON CONSTABLE – NO. 43150**
Westbound on 30th August 1958 is Ivatt Class 4MT no. 43150 which has a weekend 08.39 train from Cromer Beach to Birmingham New Street. The engine is climbing away from Melton Constable on the ex-M&GN Joint line. No. 43150 was allocated there at this time, but in the next year was transferred to Stratford and had a spell at Peterborough before condemned in January 1965. Photograph by Hugh Ballantyne courtesy Rail Photoprints.

Opposite above **MARKS TEY STATION – NO. 46467**
For medium-distance local services, a tender was desirable to increase range and reduce the need for water stops. The LMSR's strong lineage of 2-6-4T was adapted for the purpose and became a 2-6-0 designed by H.G. Ivatt. These were built between 1946 and 1953, numbering 128 examples which went on to serve across the country. No. 46467 was constructed at Darlington in June 1951 and delivered for use at Cambridge shed. The engine was there for the next ten years when transferred to the Scottish Region. No. 46467 saw further use until July 1964. Nearly new here, the locomotive is at Marks Tey station with a local train. Photograph courtesy Rail-Online.

Opposite below **MARKS TEY STATION – NO. 62795**
The 13.15 Cambridge to Colchester service exits Marks Tey station on 24th November 1952. E4 Class no. 62795 had just over two years left in traffic. Employed at Cambridge, the locomotive had been there from 1951 and arrived from Norwich. No. 62795 was amongst the class members that worked the Stainmore Line, doing so from Kirkby Stephen. Photograph by Hugh Ballantyne courtesy Rail Photoprints.

Above **MELTON CONSTABLE – NO. 64726**
Gresley J39 Class 0-6-0 no. 64726 is engaged shunting goods vans at Melton Constable on 28th August 1958. Only four months were spent at the local shed and in October the locomotive was briefly at Lowestoft before joining the ranks at Lincoln for the last two years until condemned. Photograph by Hugh Ballantyne courtesy Rail Photoprints.

Opposite **MELTON CONSTABLE SHED – NO. 65567**
The Lynn & Fakenham Railway sited their locomotive servicing facilities at Melton Constable on the southern edge of the station site. Completed in the early 1880s, a three-lane building was erected and in use to the early 1950s when demolished for a similar structure to be erected by British Railways. This stands off to the right here as Holden J17 Class 0-6-0 no. 65567 is turned on 29th August 1958. Just three months of an allocation there remained for the engine which had begun three years earlier. Standing in the background is Ivatt Class 4MT no. 43156, also allocated to Melton Constable and was there when the depot closed in 1959. The building still survives, as do the old workshop buildings just visible in the top right. Photograph by Hugh Ballantyne courtesy Rail Photoprints.

Above **MELTON CONSTABLE – NO. 61547**
Melton Constable was the point where the Lynn & Fakenham Railway diverged southward to Norwich, whilst the branch to Cromer later forged northward. The station opened in early 1882, whilst the Cromer branch was not built until the end of the decade. Holden B12 Class no. 61547 has stock ready to take the 17.48 train to Cromer at Melton Constable West Junction on 29th August 1958. The station closed in mid-1964 to passengers and the end of the year for freight. Photograph by Hugh Ballantyne courtesy Rail Photoprints.

Opposite **MILDENHALL STATION – NO. 65460**
The train commemorating the demise of the Mildenhall branch has reached the terminus behind no. 65460. According to disused-stations.org.uk, Mildenhall station consisted of station master's house, booking room, staff room, waiting room and toilet. Only one platform, 365 ft long, was provided and there was a loop for locomotives to change ends despite the provision of a turntable. A reasonably-sized goods shed was built with a 1½-ton crane, a 5-ton weighbridge and 22-cwt weighing machine. The final goods train ran on 10th July 1964. The station has since become a private dwelling. Photograph courtesy Colour-Rail.

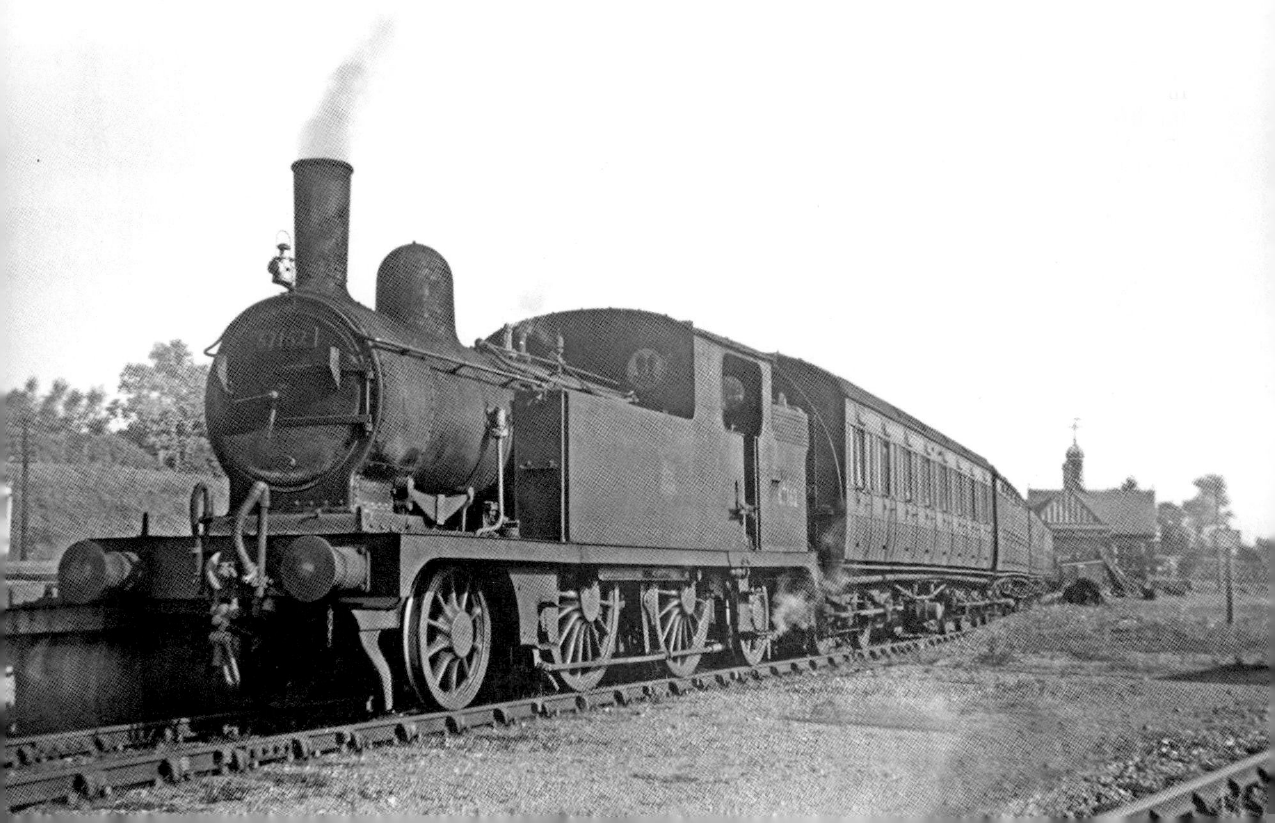

Above **NEWMARKET STATION – NO. 65438**

The Jockey Club was amongst the promoters for the Newmarket & Chesterford Railway which connected with the Liverpool Street-Cambridge main line for obvious reasons. Completed in 1848, the line ran from Great Chesterford, south of Cambridge, north-eastward to Newmarket. Yet, financial problems saw closure occur just two years later. Under new management, a proposed branch to Cambridge was seen as more desirable than the original route. Therefore, the track from Six Mile Bottom to Great Chesterford was lifted and reused to reach Cambridge. As railways in the area expanded, several stations existed at Newmarket, with the surviving structure dating from 1902. Worsdell J15 Class no. 65438 has a local train in the bay platform during the 1950s, whilst B12 no. 61567 is seen to the left. Photograph courtesy Rail-Online.

Opposite above **MUNDESLEY-ON-SEA STATION – NO. 69690**

In the late 19th and early 20th centuries, the GER stuck with 2-4-2T locomotives for their suburban services around London. A.J. Hill moved on from this in the mid-1910s by changing to the 0-6-2T wheel arrangement and the N7 (GER L77) appeared with two prototypes (saturated and superheated) followed by 20 production class members – ten of each boiler arrangement. The N7 later became a Group Standard design under the LNER and 112 were constructed to the end of the 1920s. No. 69690 was amongst 20 erected by William Beardmore & Co. in 1927, being completed during August. Shortly after Nationalisation, the engine was part of a group of class members equipped for push-pull services and allocated to the ex-GCR shed at Neasden for local branch trains. When these were discontinued, no. 69690 moved to East Anglia for the Mundesley-on-Sea to North Walsham train. This duty has taken the engine to Mundesley-on-Sea station on 31st May 1954. Photograph courtesy Rail-Online.

Opposite below **MUNDESLEY-ON-SEA STATION – NO. 67162**

The Norfolk & Suffolk Joint Railway produced two coastal lines: Lowestoft-Yarmouth; North Walsham-Cromer. The latter was completed in two stages, with the first reaching Mundesley-on-Sea in 1898 before continuing for Cromer, where a connection was made with the M&GNJR's Cromer Beach (the company promoted the N&SJR) during 1906. Mundesley-on-Sea again became the end of the line for passengers in the mid-1950s, though freight continued into the mid-1960s when the route closed completely. As mentioned above, the 2-4-2T dominated suburban traffic for the GER into the early 20th century when the 0-6-2T took over. An example of the F4 Class, no. 67162 has been displaced to the rural trains here, being at the head of the 09.00 train from Mundesley-on-Sea to North Walsham on 16th August 1951. The locomotive remained in the area until condemned on 15th August 1955. Photograph courtesy Rail-Online.

Above **NORTH WALSHAM TOWN STATION – NO. 43093**
The GER line from Norwich to Cromer was the first to pass through North Walsham and a station was built there in 1874. The Yarmouth & North Norfolk Railway gradually reached the town to establish a second station in 1881. At Grouping these were differentiated by the first becoming North Walsham Main and the ex-M&GNJR facility was named North Walsham Town. The latter stopped serving passengers in March 1959, with freight continuing to 1966. The 06.51 Peterborough to Yarmouth Beach is at North Walsham Town station on 29th August 1958. At the head of the train is Ivatt Class 4MT no. 43093. The engine was based at South Lynn to March 1959. Photograph by Hugh Ballantyne courtesy Rail Photoprints.

Opposite above **NEWMARKET YARD JUNCTION – NO. 62551**
Holden D16/3 Class no. 62551 receives the token from Newmarket Yard Junction signal box to work the 14.27 service to Ely on 3rd April 1952. Photograph courtesy Rail-Online.

Opposite below **NORTH WALSHAM STATION – NO. 65567**
To pass through North Walsham station J17 Class no. 65567 has been diverted on to the incorrect line with the 06.55 Yarmouth to Melton Constable freight train on 29th August 1958. Photograph by Hugh Ballantyne courtesy Rail Photoprints.

Below NORWICH THORPE STATION – NO. 70013

Reintroduced following the war, the 'East Anglian' train had changes made when the motive power was upgraded to the new 'Britannia' Pacifics. The departure time moved to 11.45 and the scheduled arrival at Liverpool Street was 13.55. The reverse service departed from London at 18.30, with the same 2 hours and 10 minutes allowed. Generally, the train was formed from eight vehicles. No. 70013 *Oliver Cromwell* has the southbound train departing from Norwich Thorpe station on 21st August 1951. The engine was just three months old at this point and had been delivered to Norwich shed. Photograph by T.G. Hepburn from Rail Archive Stephenson courtesy Rail-Online.

Above NORWICH THORPE STATION – NO. 61978

Even though the K3 2-6-0 was a Group Standard design, several variations and modifications were made over the construction period. The last major round of changes led to the 1302 Series built from 1934 to 1937 and formed by 74 engines erected at several locations. The alterations included roller bearings for some motion components, axlebox redesign and sanding layout changes, amongst others. No. 61978 was part of the 1302 Series, being constructed at Darlington Works in November 1936. At this time, the K3 had not been sent to the ex-GER lines and a further two years passed before they were approved. The class had been used on the M&GNJR, though the first former GER shed to take the K3s was Stratford for express goods services, but also some passenger trains. Their deployment was limited after the outbreak of war and to the end of the conflict all K3s working into the East of England were concentrated at March. No. 61978 moved there from Doncaster in October 1953 and was employed to November 1956 when transferred over to Peterborough. The locomotive has this train at Norwich Thorpe station on 17th July 1955. Photograph courtesy Rail-Online.

Above NORWICH THORPE STATION – NO. 61059

The Yarmouth & Norwich Railway built the first station at Norwich in 1844. Soon after, the connection was made with the London-Cambridge line which allowed through running between Norwich and the capital. The Eastern Union Railway arrived before the end of the decade and built Norwich Victoria station, whilst the Lynn & Fakenham Railway opened Norwich City station which later became the terminus for the M&GNJR's line from Melton Constable. Thompson B1 Class no. 61059 has an express at Norwich Thorpe station on 25th January 1958. Photograph by B.W.L. Brooksbank.

Opposite NORWICH SHED – NO. 61810

Gresley K3 Class no. 61810 is serviced at Norwich shed on 31st May 1960. The engine was March-allocated at the time. Photograph by R.C. Riley courtesy Rail-Online.

Above **OCKENDON – NO. 58062**
Johnson 1532 Class 0-4-4T no. 58062 has a local train at Ockendon on the LT&SR line around 1955. Photograph from the Dave Cobbe Collection courtesy Rail Photoprints.

Below **OCKENDON STATION – NO. 69695**
Ockendon station has a train paused at the platform with Hill N7 no. 69695 during 1957. Photograph from the Dave Cobbe Collection courtesy Rail Photoprints.

Above **PETERBOROUGH EAST STATION – NO. 69593**
Extensive sidings were present on the south side of the running lines at Peterborough East station for the transfer of freight. Displaced Gresley N2 Class 0-6-2T no. 69593 is shunting there in August 1962 – withdrawal occurred a month later. Photograph by John Briggs courtesy A1 Steam Trust.

Below **PETERBOROUGH NORTH STATION – NO. 63717**
A coal train is beginning the journey to Little Barford power station with Robinson O4 Class no. 63717 on 15th November 1958. Photograph by B.W.L. Brooksbank.

PETERBOROUGH NORTH STATION – NO. 60105
On 25th August 1962, Gresley A3 no. 60105 *Victor Wild* has an express ready to leave Peterborough. Photograph by Neville Simms from the Ranwell Collection courtesy Rail Photoprints.

Above **PETERBOROUGH NEW ENGLAND SHED – NO. 60513**
Thompson's A2/3 Class Pacifics were introduced post-war for main line mixed traffic duties. Several were based at Peterborough and saw use on named expresses, secondary passenger and freight trains. No. 60513 *Dante* was mainly at the depot from 1948 until withdrawn in 1963. Photograph by Bill Reed.

Below **PETERBOROUGH NEW ENGLAND SHED – NO. 60862**
Peterborough New England shed was home to 34 Gresley V2 Class locomotives at Nationalisation, being mainly a freight depot for the ECML. No. 60862 was there throughout the 1940s, though in 1950 rose in the ranks to join the stud at King's Cross. The locomotive is in New England yard, c. 1960. Photograph by Bill Reed.

PETERBOROUGH NORTH STATION – NO. 60125
Running through Peterborough North with the southbound 'Yorkshire Pullman' on 25th August 1962 is Peppercorn A1 Class Pacific no. 60125 *Scottish Union*. Photograph by Neville Simms from the Ranwell Collection courtesy Rail Photoprints.

Above **PETERBOROUGH NEW ENGLAND SHED – NO. 60504**
A number of depots existed at Peterborough owing to the city's importance as a junction for several lines. The GNR's first facility was close to their station and particularly large, with eight roads in the building. This was subsequently superseded by a nine-track and six-lane shed built a short distance to the north close to the goods yard. Thompson A2/2 no. 60504 *Mons Meg* is in the yard, c. 1960. Photograph by Bill Reed.

Below **PETERBOROUGH NEW ENGLAND SHED – NO. 64223**
Next to the old coal stage at Peterborough New England shed around 1960 is Ivatt J6 Class 0-6-0 no. 64223. The locomotive moved to the depot from Hornsey in February 1960 and was later condemned there in April 1961. Photograph by Bill Reed.

Above **PETERBOROUGH – NO. 60064**
After the formation of the LNER, Gresley's A1 Pacific design was chosen to work the ECML expresses and further examples were needed to assist the 12 already in service or under construction. An order for ten more was placed at Doncaster while the North British Locomotive Company was tasked with erecting another 20. No. 60064 was the first of this number and new to Edinburgh Haymarket depot in July 1924. At this time, as LNER no. 2563, the locomotive was named *William Whitelaw* to honour the former Chairman of the North British Railway and the present Chairman of the LNER. This was carried to 1941 when transferred to an A4 Pacific and no. 2563 received *Tagalie* to continue the racehorse theme of the A1/A3 Class. The engine is viewed from Spital Bridge, Peterborough, on 16th August 1958 and has a northbound express. Photograph by B.W.L. Brooksbank.

Opposite above **PETERBOROUGH NORTH STATION – NO. 69579**
Sixty-seven of Gresley's N2 Class were employed at King's Cross and Hornsey for the suburban services, whilst a number were at satellite sheds for the trains. In 1958, the process of dieselisation began and accelerated towards the early 1960s, with a number of locomotives displaced as a result. Work was found for the N2s at Peterborough or Grantham, though not enough to extend their lifespan. No. 69579 left King's Cross for Hornsey in July 1960, then for Peterborough a year later. In September 1962, the engine was sent for scrap. This date was just a month away here as the engine stands against the platform at Peterborough North. Photograph by B.W.L. Brooksbank.

Opposite below **PETERBOROUGH NORTH STATION – NO. 67398**
Empty coaching stock is moved by Ivatt C12 Class 4-4-2T no. 67398 at Peterborough North station on 20th July 1957. Joining the local depot at the start of the year, the locomotive worked there to November 1958 when condemned. Photograph by B.W.L. Brooksbank.

Below **PETERBOROUGH NORTH STATION – NO. 60021**

The premier train on the ECML was the 'Flying Scotsman'. In 1927, the service was afforded even greater prestige with the journey being made non-stop in the summer months. Gresley's A1/A3 Pacifics dominated this to the mid-1930s when the A4 Streamlined Pacifics were introduced. After the war, the 'Flying Scotsman' lost the non-stop schedule to a new train – 'The Capitals Limited'. In 1953, the name became 'The Elizabethan' to honour the accession of Queen Elizabeth II. Six-and-three-quarter hours were allowed for the ten-coach train to travel between King's Cross and Edinburgh, with the A4s again at the head. On 19th July 1958, no. 60021 *Wild Swan* has the northbound train at Peterborough station. The engine was one of nine A4s used on 'The Elizabethan' during the season, recording 12 journeys between the capitals. Photograph by D.C. Ovenden courtesy Colour-Rail.

Above PETERBOROUGH NORTH STATION – NO. 60108

Another locomotive recorded passing through Peterborough North station on 19th July 1958 was Gresley A3 Class Pacific no. 60108 *Gay Crusader*, which has a King's Cross to Leeds and Bradford express. The engine was amongst ten ordered by the GNR shortly before Grouping and appeared from Doncaster Works in June 1923. The locomotive was new to the town's shed and was employed there several times until condemned in October 1963. Over this period, no. 60108 was mainly on the southern half of the ECML, but did have spells on the ex-Great Central main line at Gorton and Neasden in the 1940s and 1950s. When pictured, *Gay Crusader* was serving at King's Cross depot. Crescent Bridge is present in the background and this carries a road over the running lines. The structure was built in 1913 and replaced a level crossing. Photograph by D.C. Ovenden courtesy Colour-Rail.

Above **PETERBOROUGH**
View south from Westwood Bridge, Peterborough, towards Peterborough North station, which lies behind Spital Bridge in the distance. A Gresley V2 Class 2-6-2 is heading northward with a King's Cross to Newcastle express on 18th August 1962. Photograph by B.W.L. Brooksbank.

Opposite above **PETERBOROUGH NORTH STATION – NO. 92041**
Much of the ECML freight was concentrated at Peterborough for dispatch elsewhere. Coal traffic from Yorkshire arrived at Peterborough for transfer to London and this task was performed by a succession of locomotives over the years, with the last to take the role being the BR Standard Class 9F 2-10-0. No. 92041 of the class has such a train here at Peterborough North station on 18th August 1962, being viewed from Crescent Bridge looking to the station, with the GNR hotel on the right. The engine had a year left at Peterborough, having been allocated there from new in late 1954, and was at Colwick, Barrow Hill and Langwith Junction before condemned during 1965. Photograph by B.W.L. Brooksbank.

Opposite below **PETERBOROUGH NORTH STATION – NO. 76086**
The railway was well established in Peterborough before the arrival of the GNR. The London & Birmingham Railway had branched from their main line, as had the Eastern Counties Railway, whilst the Midland Railway also built a short connection from Syston, north of Leicester. When the GNR's 'Lincolnshire Loop' was ready, the company initially used the ECR's station until 1850 when the new main line was ready. At Grouping, the two became Peterborough East and Peterborough North respectively. BR Standard Class 4 2-6-0 no. 76086 is at the latter with a Leicester to Peterborough local train on 12th April 1958. Peterborough East closed in 1966 and North was extensively rebuilt in the 1970s. Photograph by D.C. Ovenden courtesy Colour-Rail.

Above **POTTERS BAR – NO. 60123**

The 'West Riding Limited' was one of the trio of streamlined trains introduced by the LNER in the mid-1930s. All were withdrawn during the war and returned by varying degrees afterwards. The 'West Riding' appeared in 1949 to serve travellers between King's Cross, Leeds and Bradford. The southbound train left Leeds at 07.50 and had a four-hour schedule, whilst northward departure from King's Cross occurred at 15.45, with a time just over three hours fifty minutes. Some of the coaching stock from the 'West Riding Limited' was reused by BR and at least an articulated twin is behind the tender of Peppercorn A1 Class Pacific no. 60123 *H.A. Ivatt*. The locomotive has the southbound train at Potters Bar on 8th May 1954, being allocated to Ardsley shed, Leeds, at the time. Photograph by George C. Lander courtesy Rail Photoprints.

Below **POTTERS BAR STATION – NO. 60010**
The original station at Potters Bar still stands here in the early 1950s, as Gresley A4 Pacific no. 60010 *Dominion of Canada* passes through with an express. In use from August 1850, the station became a particular bottleneck as the four lines merged to just two there. Under BR, Potters Bar was completely rebuilt with a 'modern' styling and two island platforms to allow expresses uninterrupted running. Photograph courtesy Rail-Online.

Below REEDHAM STATION – NO. 61971

Reedham station opened in 1844 on the first railway in Norfolk – the Yarmouth & Norwich. A point on this line just to the south of Reedham was chosen as the junction for the route from Lowestoft and this was operational from 1847. In the early 20th century, Reedham was rebuilt on a new site closer to the Lowestoft junction and this station continues to serve the village. Gresley K3 Class no. 61971 has an express at Reedham in October 1960. For a number of years a Norwich engine, no. 61971 was sent to Colwick in early 1961, though was condemned for scrap just two months later. Photograph courtesy Rail-Online.

Above POTTERS BAR – NO. 61623

With the authorisation of the GNR main line a number of satellite schemes to make a connection also came to fruition. One was the Hitchin & Royston Railway which ran from the GNR route to Shepreth, south west of Cambridge, to meet the Eastern Counties Railway. The first section was ready in 1850 and trains to Cambridge ran the following year. The GNR initially let the ECR lease the line and passengers changed at Hitchin, yet at the end of the term in the 1860s, the GNR obtained full running powers to Cambridge. The services were allowed to become secondary considerations as time progressed before an attempt to revitalise these was made in the early 1930s. A fast-paced train with buffet carriage in the formation appeared in 1932 and was soon popular with travellers. A mixture of locomotives and coaching stock were used on the route to just before the war when around eight of Gresley's B17s were grouped at Cambridge for the services to King's Cross. No. 61623 *Lambton Castle* transferred there in October 1945 and remained employed until withdrawn during July 1959. The locomotive has a southbound Cambridge-King's Cross express at Potters Bar on 8th May 1954. Photograph by George C. Lander courtesy Rail Photoprints.

Above SANDY STATION – NO. 60028

Around halfway between Hitchin and Huntingdon, Sandy station opened in 1850. A second station was built in the 1850s on the Bedford-Cambridge line and the pair co-existed to 1968 when the ex-GNR station became the sole one serving the town. Like Potters Bar, Sandy station was a two-track bottleneck and in the 1970s, the opportunity was taken to rebuild the facility. Gresley A4 Pacific no. 60028 *Walter K. Whigham* is on the main line with an express in June 1962. The engine was one of the top performers at King's Cross during the 1950s, though was condemned there at the end of 1962. Photograph courtesy Colour-Rail.

Opposite above REEDHAM STATION – NO. 67704

Whilst the London Midland & Scottish Railway introduced a large number of 2-6-4T engines for suburban and short intercity duties, the LNER only built one, relying on the 0-6-2T type. Edward Thompson decided to move over to the 2-6-4T during the war and the prototype L1 Class engine, no. 9000, entered traffic in 1945. The engine was part of an initial order for 30, though the remainder did not appear until after Nationalisation. A further 70 were built through to 1950. Around half of the first 30 constructed were allocated to Stratford shed and this included no. 67704. The engine went on to work at several locations across Eastern England and had reached Lowestoft when captured here at Reedham station in July 1956. No. 67704 has a Yarmouth to Norwich train. Photograph courtesy Rail-Online.

Opposite below SANDY STATION – NO. 60055

Gresley A3 Pacific no. 60055 *Woolwinder* has an express at Sandy station during 1956. In June of that year, the locomotive transferred from Doncaster to King's Cross, which proved the final allocation. No. 60055 was employed there until condemned in September 1961. Photograph courtesy Rail-Online.

Above SANDY STATION – NO. 48550

Stanier 8F Class 2-8-0 no. 48550 has a freight train – with a number of coal wagons – on the old LNWR side of Sandy station on 28th March 1959. During the Second World War, the railways were Nationalised and operated by the Railway Executive Committee. The latter placed orders for the good of the war effort and this included Stanier 8F Class 2-8-0s to be used for the movement of goods and war materials. The LNER was obliged to build class members at both Doncaster and Darlington, with no. 48550 amongst 30 completed by the latter. The engine was then loaned to the LNER for use. After the end of the conflict, the 8Fs began to be repatriated to the LMSR and no. 48550 was at Northampton for Nationalisation in 1948. In late 1950, a move was made to Bletchley and the locomotive was there for the next 15 years. Photograph by G.H. Hunt courtesy Colour-Rail.

Opposite SANDY – NO. 45147

A branch from the London & Birmingham Railway – subsequently London & North Western Railway – at Bletchley reached Bedford in 1846. In the late 1850s, the Sandy & Potton Railway was built by a local landowner and became part of the Bedford & Cambridge Railway which reached completion in the early 1860s. The two above-mentioned sections joined with others to form the 'Varsity Line' from Oxford to Cambridge which was a key cross-country route. From Bedford to Sandy the branch was originally a single line which crossed the ECML north of the station and ran down to the eastern side, then continuing southward to the junction for Cambridge. Bletchley-allocated Stanier Class 5 4-6-0 no. 45147 appears to be coming off the flyover here with a local freight on 22nd December 1962. Originally an iron girder bridge, this was replaced under BR by a steel girder structure, yet this was not enough to prevent the line's closure in 1968. The engine arrived at Bletchley shed in November 1961 and remained on the roster there to March 1964 when a transfer to Stafford occurred. Withdrawal from Aintree took place during May 1967. Photograph by Neville Simms from the Ranwell Collection courtesy Rail Photoprints.

Above **SAXMUNDHAM STATION – NO. 70002**

Between Ipswich and Lowestoft, Saxmundham station was built in 1859. At the same time, the first section of the Aldeburgh branch was ready, with the second completed in early 1860. Passenger services from Saxmundham to Aldeburgh lasted until 1966, though freight continues to run to just north of Aldeburgh owing to the presence of the power-generating facility at Sizewell. In 2018, Saxmundham station was the scene of a fire which required extensive renovations costing £1,000,000. These saw the original two-storey building reduced to a single-storey structure with improved facilities. Travelling in the direction of Ipswich on 11th September 1960 is BR Standard Class 'Britannia' Pacific no. 70002 *Geoffrey Chaucer*. The locomotive was in the middle of a two-year allocation to Norwich and at the end was sent on to March depot. Photograph courtesy Colour-Rail.

Opposite above **SAXMUNDHAM STATION – NO. 67709**

View southward to the level crossing at Saxmundham station on 11th October 1958. Thompson L1 Class no. 67709 has taken water, whilst a DMU working on the Aldeburgh branch stands off to the left. The engine had recently broken a spell of seven years at Ipswich shed and relocated to Lowestoft in September 1957 for an 18-month sojourn. The final move to Stratford occurred in January 1959 and no. 67709 was condemned there during December 1960 after a service life lasting less than 13 years. Photograph courtesy Rail-Online.

Opposite below **SAXMUNDHAM STATION – NO. 65447**

In April 1956, Worsdell J15 Class 0-6-0 no. 65447 departs from Saxmundham station with a train for Aldeburgh. Built in August 1899 at Stratford, the locomotive was in traffic until April 1959. For many years, the engine was Ipswich-allocated. Photograph courtesy Rail-Online.

Below SHENFIELD STATION – NO. 70000

The Great Depression brought about a shift in thinking for economists. Before the crash, the typical response would have allowed the markets to correct themselves naturally, whereas in the early 1930s intervention by governments was thought preferable to lessen the impacts and duration of the recession. This action was fully embraced in America, whilst the British Government adopted a more cautious approach. Money was made available at low interest to companies for works that created jobs and improved public services. The railways quickly devised schemes to take advantage of the terms, with several of the LNER's receiving funding. One of these was the electrification of the line from Liverpool Street station to Shenfield & Hutton station. By 1939, two miles of line had overhead gantries ready for lines to be hung, yet the outbreak of war quickly brought a halt to work. Activities resumed post-1945 and the 1500V DC system was ready for trains in 1949. On 17th July 1951, BR Standard Class 'Britannia' Pacific no. 70000 *Britannia* has a Norwich to Liverpool Street express under the wires at Shenfield station. Photograph by George C. Lander courtesy Rail Photoprints.

Above SHENFIELD – NO. 61201

Almost 21 miles from London, the village – later town – of Shenfield was skirted by the Eastern Counties Railway's Brentwood to Colchester line. A stop was later provided for the area in 1847, though this was discontinued some three years later. In early 1887, a new station was opened in preparation for the completion of two new branch lines, one to Southend and another to Southminster. The new facility was called Shenfield & Hutton Junction, later just Shenfield & Hutton which persisted until 1968. Thompson B1 no. 61201 is at Shenfield with an express in 1952. The locomotive was amongst 150 class members ordered from the North British Locomotive Company in 1946, emerging from Queen's Park Works in June 1947. Initially allocated to Doncaster, the engine was employed at Ipswich between 1950 and 1953, thereafter working on the ex-GCR lines for the remainder of the decade. Photograph courtesy Rail Photoprints.

Above SOUTH LYNN STATION – NO. 61545
Holden B12 Class 4-6-0 no. 61545 has the 13.45 Birmingham to Yarmouth train at South Lynn station in July 1956. Throughout the 1950s, the locomotive was mainly at one of the sheds in Yarmouth, though there were some spells at Norwich. No. 61545 was condemned early in 1957. Photograph courtesy Rail-Online.

Opposite above SHENFIELD – NO. 61332
Before the introduction of the 'Britannia' Class Pacifics, Thompson's B1s worked the top expresses on the Great Eastern Main Line. No. 61332 has the London-bound 'East Anglian' at Shenfield on 18th June 1949 when approaching a year old. New to Cambridge, within weeks the locomotive had moved to Norwich and held employment there over five years. No. 61332 was then transferred to the Scottish Region and had nine years at Edinburgh St Margaret's shed. When pictured, the overhead lines at Shenfield were not quite ready and electric services did not start until 26th September. Photograph by B.W.L. Brooksbank.

Opposite below SNETTISHAM STATION – NO. 62510
A Hunstanton-Cambridge local train is at Snettisham station during the mid-1950s. The village was just a few miles south of Hunstanton and the station was on the Lynn & Hunstanton Railway which began operations in 1862. This was later purchased by the GER in 1890. The locomotive – with safety valves lifting – is Holden D16/3 Class no. 62510. Allocated to Melton Constable in 1955, the engine moved to Cambridge in 1956 and was condemned at King's Lynn in 1957. Snettisham station was open to May 1969 and the building has been converted into a home. Photograph courtesy Rail-Online.

Above SOUTH LYNN STATION – NO. 67386
As mentioned, Ivatt's C12 Class 4-4-2T were displaced from their original sphere of operations in London and Yorkshire to find employment elsewhere. No. 67386 moved from Leeds to King's Lynn in the early 1950s for use on the short distance trains between the latter and South Lynn. Pictured at South Lynn in 1955, the engine saw another three years' service before withdrawn. Photograph courtesy Rail-Online.

Opposite SOUTH LYNN STATION – NO. 61530
The relatively short Lynn & Sutton Bridge Railway opened between the two places in 1866, though the original station was called West Lynn. During 1886, this closed and was replaced by South Lynn station which served passengers to 1959. Holden B12 Class no. 61530 is with a local train at South Lynn during the early 1950s. The locomotive was new in 1914 and served up to November 1959. In 1951 no. 61530 was at Norwich then Yarmouth and continued to switch between the two up to 1959 when six months at Cambridge took the locomotive to the withdrawal date. Photograph courtesy Rail-Online.

Above **SOUTH LYNN STATION – NO. 43095**
In 1950/1951, a number of Ivatt Class 4MTs were scheduled to begin work in the Scottish Area. These plans soon changed to take the engines on to the ex-M&GNJR and dispatch Gresley K2s northward. Nos 43090-43095 were new to the shed at South Lynn. This was an important point on the line as trains mostly changed engines there, with through-working quite rare. No. 43095 arrived at South Lynn in December 1950 and was at the depot until closure of the line in 1959. This event was around six months away here as the locomotive has taken over on the Lowestoft to Nottingham and Derby express on 30th August 1958. Incredibly, the Eastern Region rebuilt South Lynn shed during the year and ceased operations in 1959. No. 43095 continued work, briefly at Boston before joining the ranks at Lincoln for several years. Withdrawal occurred in November 1966 after two years at Carnforth. Photograph by Hugh Ballantyne courtesy Rail Photoprints.

Below SOUTH LYNN STATION – NO. 67374
Taking water from a platform column at South Lynn station is Ivatt C12 Class 4-4-2T no. 67374. The locomotive was briefly allocated to South Lynn shed in mid-1951 before transferring to the nearby King's Lynn depot a short time later. No. 67374 remained there until condemned during April 1958. On the other side of the platform is Ivatt 4MT no. 43095. The engine was built new with a tablet catcher in order to work on the M&GNJR. Another feature when built was the application of unlined black livery which was done by both Doncaster and Darlington Works when they contributed class members. The first BR emblem is also of the smaller style. Photograph courtesy Rail-Online.

Above ST IVES STATION – NO. 65562

The Railway Correspondence & Travel Society organised 'The Fensman' railtour for 24th July 1955. This ran from Liverpool Street to Cambridge behind 'Britannia' Pacific no. 70037 *Hereward the Wake*, then Worsdell J15 Class no. 65562 took over to take the party on several branches in the area. The train has reached St Ives station here which was east of Huntingdon and a hub for several local lines. The station was opened on 17th August 1847 on the line from Cambridge to Huntingdon. A connection with Ely was made from St Ives in the late 1870s, whilst a line from this also reached March and Ramsey. The latter was a destination for the tour from St Ives, which closed in 1970. Photograph courtesy Rail Photoprints.

Opposite SOUTHEND-ON-SEA (VICTORIA) STATION – NO. 61603

Edward Thompson rebuilt ten of Gresley's B17s. These featured a B1 Class boiler and had the number of cylinders reduced from three to two, becoming B2 Class 4-6-0s. In trials these were found to be inferior to Thompson's own B1s, as well as B17s that retained three cylinders, though used the B1 boiler, which worked at 225 lb per sq. in. (200 for the original boiler). No. 61603 *Framlingham* was an early rebuild in October 1946. The engine is arriving at Southend-on-Sea (Victoria) station with a service from Shenfield on 18th April 1954. The London, Tilbury & Southend Railway first served these areas and 30 years elapsed before the GER sought to take some of the traffic for itself. Their route left the main line at Shenfield and headed eastward to the north of the LT&SR's lines. A branch was also constructed to Southminster. The GER's station at Southend-on-Sea opened from October 1889 and after Nationalisation had Victoria added to differentiate from the LT&SR station which became Central. Photograph courtesy Rail-Online.

Above ST NEOTS STATION
A quiet scene at St Neots station, taken on 17th May 1956. Located between Sandy and Huntingdon, the station opened on 7th August 1850 and continues to serve passengers for the town. Photograph by B.W.L. Brooksbank.

Opposite above STOKE STATION – NO. 62510
Between Sturmer and Clare on the Stour Valley Railway, Stoke station opened on 9th August 1865. Holden D16/3 Class no. 62510 has a local train there on 24th September 1956. From Bury St Edmunds to Cambridge, no. 62510 had recently transferred to the latter's depot and was employed there to January 1957. Moving to King's Lynn, the locomotive did not survive past the end of the year. Stoke station was in use to 1967. Photograph from the Dave Cobbe Collection courtesy Rail Photoprints.

Opposite below STRATFORD SHED – NO. 70000
At the height of operations, Stratford shed had over 500 locomotives on the roster for all types of duties in the London area and on the main line. This number reduced slightly at Nationalisation and the types had been partially modernised under the LNER. One of the first of BR's new Standard Classes at Stratford was Class 7 Pacific no. 70000 *Britannia* when completed at Crewe Works in January 1951. The engine went on to work there to January 1959. Eighteen other class members were at Stratford for varying periods over the years. The only other Standard Class to be at the depot for any length of time was the Class 4MT 2-6-0, though some others were briefly trialled under BR. In the background to the right is the shed's coal stage, which was one of the largest in the country and installed on the site during 1915. Photograph by Bill Reed.

Above **STRATFORD SHED – NO. 69709**

Hill N7 Class 0-6-2T no. 69709 appears to have been stopped for attention at Stratford shed in the late 1950s. Hanging from the handrail is a chain winch which has likely been used to remove part of the motion, as a component of this sits on the running plate. The locomotive was amongst 32 ordered by the LNER from Doncaster Works and appeared from there in December 1927. From this time, the engine retains the Westinghouse pump, though has lost the condensing apparatus, whilst the round-top firebox was part of the design from new. No. 69709 was at Stratford shed for Nationalisation (along with most other class members) but was transferred to Norwich in 1949 as Thompson L1 Class engines took over the N7 duties. No. 69709 later returned to Stratford in October 1957 and stayed to withdrawal in November 1960. Photograph by Bill Reed.

Opposite above **STRATFORD – NO. 61572**

The first Holden B12 4-6-0 was withdrawn from service in 1945 and the class total of 80 steadily diminished through the 1950s. The end of the decade saw the most engines condemned, with 17 in 1957 and 15 in 1959. At this time, just one remained in traffic – no. 61572. The locomotive was employed at Norwich on both passenger and goods trains to 1961 when finally taken out of use. Being the last B12, the opportunity to preserve the locomotive was taken. Following a period of storage at Stratford – where the engine is seen here on 17th August 1962 – the Midland & Great Northern Joint Railway Society purchased no. 61572 which was later transported to the North Norfolk Railway. After an extensive restoration, the locomotive returned to steam in 1995 and has been operational for various periods since. At the time of writing, no. 61572 is under renovation. Photograph by D.J. Dippie.

Opposite below **STRATFORD SHED – NO. 65469**

Several locomotives are at Stratford shed on 17th August 1962. The first shed in the area was opened in 1840 and several buildings were erected for stabling subsequently. The ones in use for the most time occupied land on the west side of the line to Cambridge, being constructed in 1871 and 1887. Here, Worsdell J15 Class no. 65469 has reached Stratford for scrapping following withdrawal from March at the start of the month. Hill J19 Class 0-6-0 no. 64673 is behind and also recently out of traffic. Thompson L1 no. 67730 – on the extreme right – was still in steam at the time, yet did not see out the month before marked for scrap. Steam stabling at Stratford had all but ceased by the start of 1963. Photograph by D.J. Dippie.

Above STRATFORD MARKET STATION – NO. 69714

In the mid-1840s, the Eastern Counties & Thames Junction Railway was promoted to connect the company from the main line at Stratford with the docks to the south and south east. Stratford Bridge station was built with this route in 1847. Subsequently part of the ECR and GER, the station was close to the GER's Stratford Market, which was constructed in the late 1870s as a rival to Spitalfields. Stratford Bridge was renamed Stratford Market as a result. Despite the proximity of the market, traffic at the station became light in the 1950s and was discontinued before the end of the decade. The last train is seen on 4th May 1957 with Hill N7 Class no. 69714. With the lines remaining active, development of the area in the 1990s and 2000s saw the station revived as part of the Jubilee line and Docklands Light Railway. Photograph by M.J. Reade courtesy Colour-Rail.

Below **STRATFORD STATION – NO. 47300**
On Saturday, 14th April 1951, the Railway Correspondence & Travel Society took members on the 'East London' railtour. This began at Fenchurch Street and headed to North Woolwich, returned to Stepney, then went to East Ham. The journey ended at Stratford with Fowler 3F Class 0-6-0T no. 43700. Photograph by B.W.L. Brooksbank.

Below **STRATFORD WORKS – NO. 62553**
A perhaps unrecognised role for many railway workshops was the dismantling of life-expired engines. This had been done quietly at the respective sites before the end of steam was announced by BR in the mid-1950s. Thereafter, these locations became particularly interesting for enthusiasts. At Stratford Works, several shops were employed at various times, with Holden D16/3 no. 62553 in the Stripping Shop here on 27th January 1957. The locomotive had been condemned at the start of the year, whilst Worsdell J15 Class no. 65404, which is behind, left traffic in October 1956. As steam came to an end on ex-GER lines relatively early, Stratford was busy scrapping locomotives to late 1961. Photograph courtesy Rail-Online.

Above **STRATFORD LOW LEVEL STATION – NO. 68971**
A connection between the London & North Western Railway and the docks at Poplar was promoted in the late 1840s and ready (as the North London Railway) for the start of the new decade. This line was a short distance to the west of Stratford and a connection from there was soon deemed desirable. This ran from Victoria Park and opened to traffic in 1854. To reduce congestion at Stratford main line station, Stratford Low Level station opened to serve traffic on the NLR. A freight train is passing through in October 1955 behind Gresley J50 0-6-0T no. 68971. At this time the engine was Hornsey-based. Photograph by F. Hornby courtesy Colour-Rail.

Above SWAFFHAM STATION – NO. 62545
A Norwich to King's Lynn local train arrives at Swaffham station on 18th September 1952. The locomotive is Holden D16/3 Class 4-4-0 no. 62545 which was part of the original D15 Class order from Stratford in January 1904. Conversion to D16/3 specifications occurred in March 1933 and the engine ran in this form to withdrawal during September 1958. When pictured, no. 62545 was on the cusp of a transfer from Lowestoft to Norwich, having been apparently on loan for the summer season. Before leaving traffic, the engine went on to work at Bury St Edmunds, Cambridge and King's Lynn. Photograph by George C. Lander courtesy Rail Photoprints.

Below **SUDBURY STATION – NO. 62790**

The original terminus for the branch from Marks Tey, which was opened by the Colchester, Stour Valley, Sudbury & Halstead Railway, was at Sudbury. In use from late July 1849, the facility was later moved to the west when the line was developed to reach Cambridge. At this time, the original station changed use to just goods and freight. Holden E4 Class no. 62790 has a Cambridge to Colchester train ready to leave Sudbury in the early to mid-1950s. Transferring from King's Lynn to Cambridge in June 1952, the locomotive was condemned at the latter in January 1956. Sudbury has once again become the terminus of the line from Marks Tey following the closure of the Cambridge extension in 1967. As the facilities were seriously reduced towards the new millennium, a new station was constructed in the early 1990s. Photograph courtesy Rail-Online.

Above SWAFFHAM STATION – NO. 62608

On the Lynn & Dereham Railway, Swaffham station served passengers from 10th August 1847. A short distance to the east, Swaffham became the point of connection for the Watton & Swaffham Railway project which was part of the Thetford & Watton Railway scheme. The latter left the Norwich & Brandon Railway east of Thetford and was ready first in 1869, whilst the second half followed in 1875. Here, in July 1955, Holden D16/3 Class no. 62608 has a King's Lynn to Dereham local train exchanging the line token with Swaffham signal box. This was at the east end of the station, next to the level crossing, which has a bus waiting (visible on the right). Swaffham station closed in September 1968. Photograph courtesy Rail-Online.

Opposite THETFORD STATION – NO. 70003

The Great Eastern Main Line was one of the first to dispense with steam traction. To mark this solemn occasion, the Railway Correspondence & Travel Society organised a railtour for 31st March 1962. Starting at Liverpool Street, the train ran via Ipswich to Norwich behind BR 'Britannia' Pacific no. 70003 *John Bunyan*, then changing over to Holden J17 Class no. 65567 to take the group via Dereham to Foulsham, running back via Swaffham to Thetford. No. 70003 returned to the head of the train and travelled to Cambridge before running onwards to Liverpool Street. The changeover of engines is witnessed here by an enthralled group of spectators at Thetford station. Photograph courtesy Rail-Online.

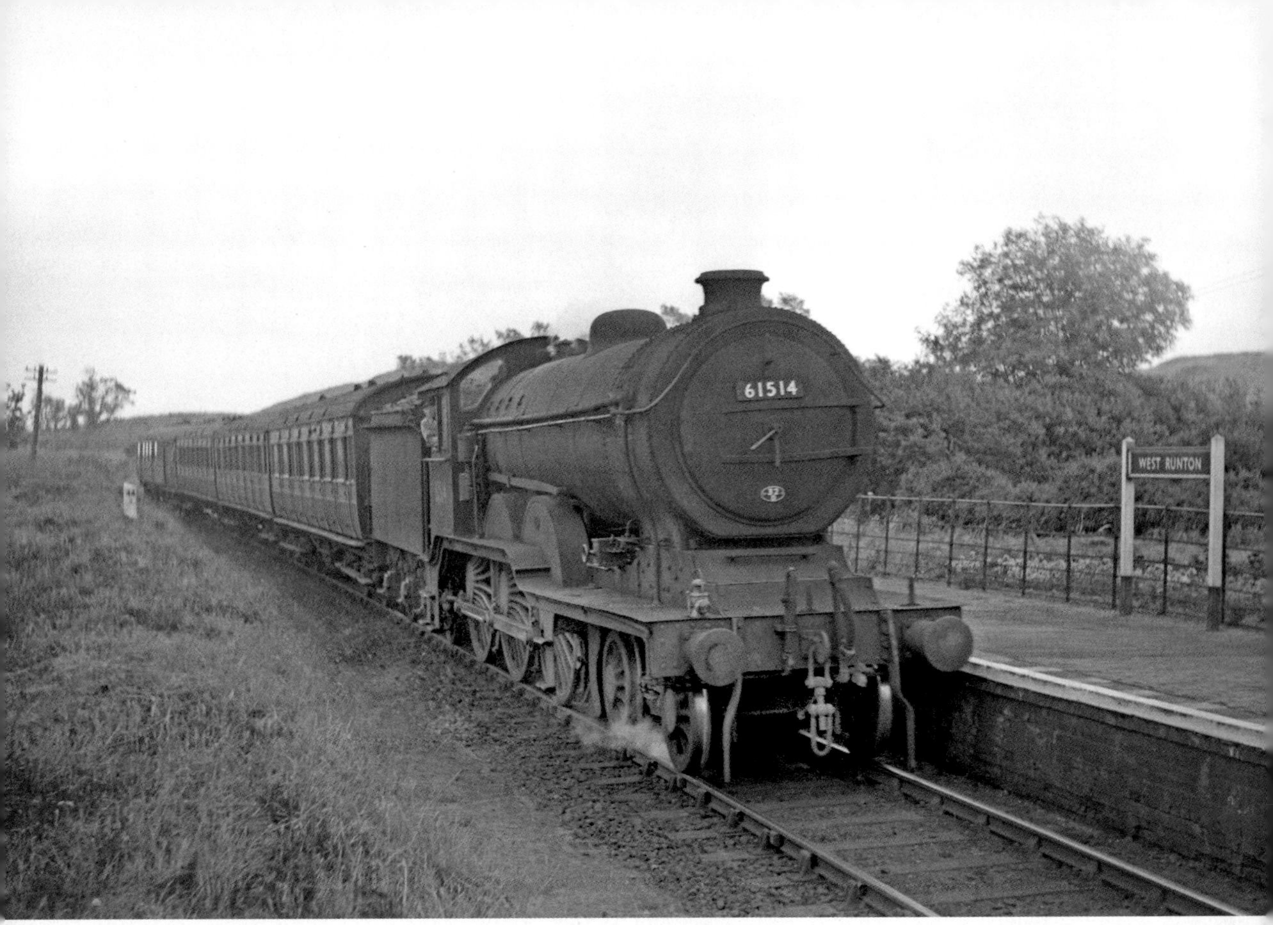
61514
WEST RUNTON

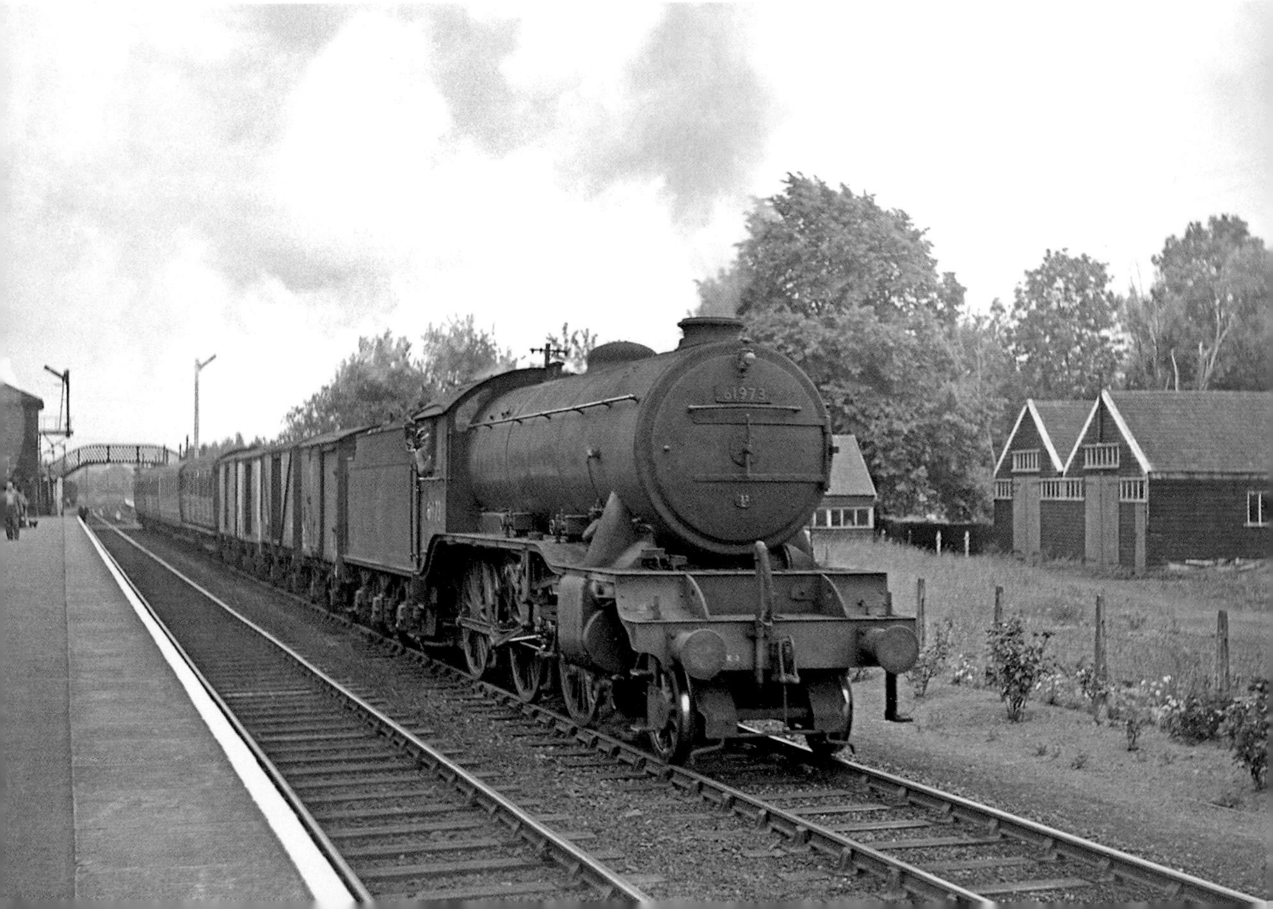
61973

Above **WITHAM STATION – NO. 67191**
Between Chelmsford and Colchester on the main line, Witham station dates from 1843. Later, branches connected to Braintree and Maldon. Holden F4 Class no. 67191 has a local destined for the latter in 1954. The engine was condemned at the end of the following year. Photograph from the John Day Collection courtesy Rail Photoprints.

Opposite above **WEST RUNTON STATION – NO. 61514**
Around two miles west of Cromer, West Runton station began serving the local village in 1887. Despite many local closures, West Runton continues to operate, whilst Sheringham station, which was the next to the west, has become part of the heritage North Norfolk Railway. Holden B12 no. 61514 has a southbound local train from Cromer to Norwich at West Runton during July 1956. The engine was Norwich-allocated from July 1955 to October 1959. Photograph courtesy Rail-Online.

Opposite below **WHITLINGHAM STATION – NO. 61973**
A local train from Norwich to Lowestoft also features a small number of empty fish vans ready to transport the next catch inland during July 1956. At the head of the formation is Gresley K3 Class no. 61973 and the train is at Whitlingham station. The Yarmouth-Norwich line had been in use for 30 years before Whitlingham was added to the stations along the route. Three miles east of Norwich, the facility came into use at the same time as the line from Norwich to North Walsham and later Cromer – initially known as Whitlingham Junction – being to the west of the connection. Despite both lines continuing to operate, Whitlingham was closed in September 1955. For over ten years after Nationalisation, the locomotive was Lowestoft-based. In November 1962, no. 61973 was condemned at Doncaster. Photograph courtesy Rail-Online.

Above **WHITLINGHAM STATION – NO. 62592**
Another scene captured at Whitlingham station shortly before closure – 17th July 1955 – and featuring Holden D16/3 no. 62592, which is travelling towards Norwich. The line to North Walsham and Cromer can be seen curving away in the background on the left. With Norwich shed's '32A' shed code on the smokebox door, the locomotive worked from there over several occasions in the 1950s. The depot always had a large number (generally at or above 20) for services in the area. No. 62592 was withdrawn from King's Lynn in April 1958. Photograph courtesy Rail-Online.

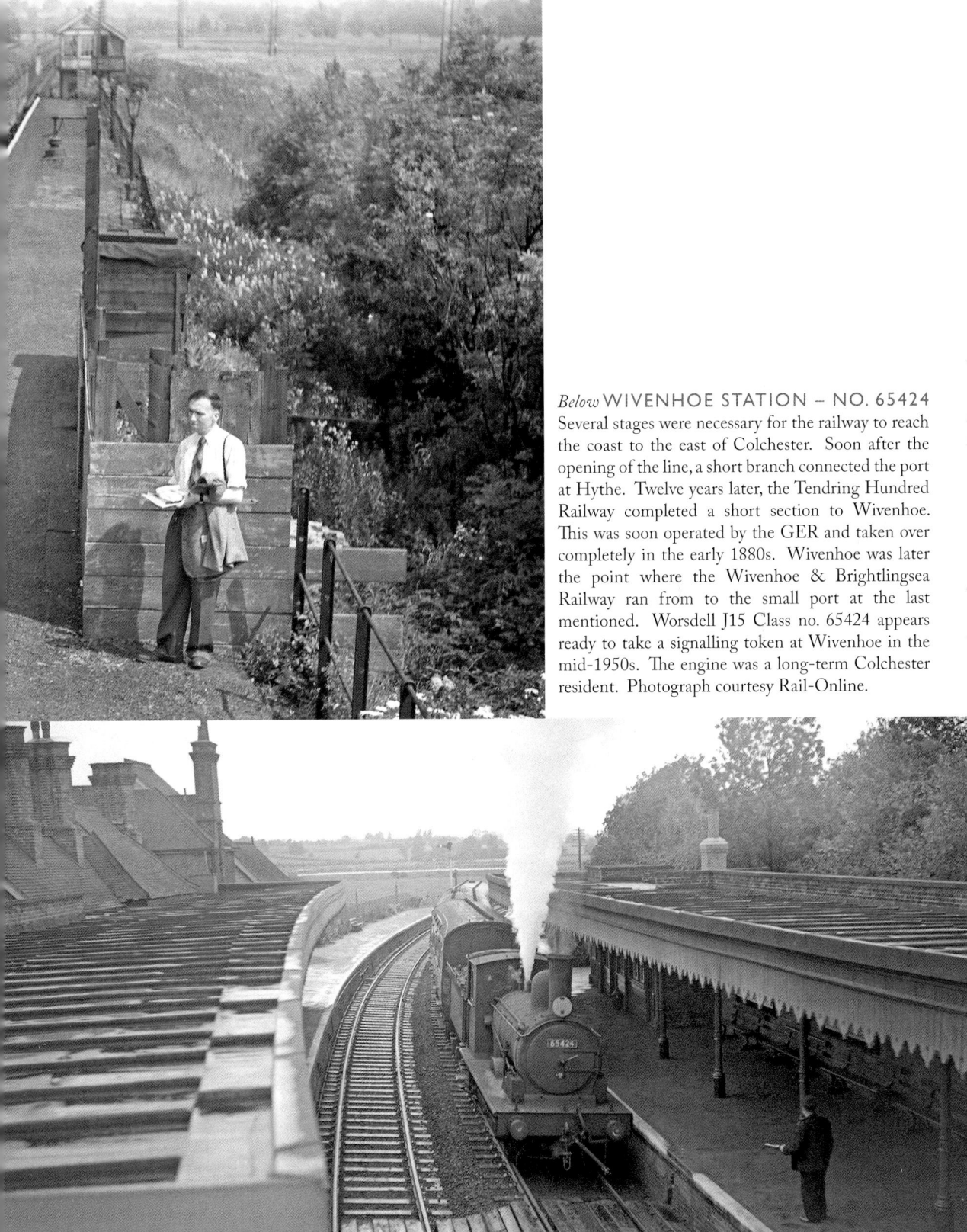

Below WIVENHOE STATION – NO. 65424
Several stages were necessary for the railway to reach the coast to the east of Colchester. Soon after the opening of the line, a short branch connected the port at Hythe. Twelve years later, the Tendring Hundred Railway completed a short section to Wivenhoe. This was soon operated by the GER and taken over completely in the early 1880s. Wivenhoe was later the point where the Wivenhoe & Brightlingsea Railway ran from to the small port at the last mentioned. Worsdell J15 Class no. 65424 appears ready to take a signalling token at Wivenhoe in the mid-1950s. The engine was a long-term Colchester resident. Photograph courtesy Rail-Online.

Above YARMOUTH BEACH STATION – NO. 68651

Three batches of A.J. Hill's J68 (GER C72) 0-6-0T were ordered from Stratford Works over 11 years, starting in 1912. No. 68651 was part of the second order consisting of ten engines in 1913/1914. Mainly used for shunting duties, no. 68651 was attached to Yarmouth Beach from the mid-1940s to withdrawal in May 1958. The engine is seen between the station and shed yard on 14th July 1951. Also visible is Ivatt Class 4MT no. 43104 which was new to South Lynn just four months earlier. Holden J17 Class no. 65581 stands to the right and was another Yarmouth engine. Photograph by B.W.L. Brooksbank.

Opposite WYMONDHAM STATION – NO. 62561

To the west of Norwich, Wymondham station was ready for traffic with the Norwich & Brandon Railway in July 1845. The station was soon upgraded to a junction with the opening of the Dereham branch during 1847. A connection between this and King's Lynn was available in the following year. Wells-next-the-Sea was reached by an extension via Fakenham from 1857. The Ipswich-Norwich line passed by Wymondham a short distance to the south east and the GER resolved to construct a short branch in the early 1880s. Holden D16/3 Class no. 62561 has a train from Norwich to Wells at Wymondham during the late 1950s. An early DMU with a Wells destination board is on the opposite line. The branch to Wells closed in the late 1960s and a section going past Dereham has since reopened at the Mid-Norfolk Railway. Photograph courtesy Rail-Online.

Above YARMOUTH BEACH STATION – NO. 43108

The 18.12 train to South Lynn departs Yarmouth Beach station on 28th August 1958. The locomotive is Ivatt Class 4MT no. 43108 and was new to the first mentioned in May 1951. Recorded as working on the last day of the M&GNJR, the engine had several allocations subsequently, first going to Boston, then Colwick six months later. In the mid-1960s, no. 43108 was at Peterborough, Staveley, Canklow and Langwith Junction. Withdrawal occurred in November 1965. Photograph by Hugh Ballantyne courtesy Rail Photoprints.

Below **YARMOUTH BEACH STATION – NO. 62517**
Originally the last station on a short line from Yarmouth to Martham, Yarmouth Beach started life as Yarmouth on 7th August 1877 under the Great Yarmouth & Stalham Light Railway. This later became the Yarmouth & Norwich Light Railway and an extension was built to North Walsham. Yarmouth Beach ultimately became the terminus for the M&GNJR when amalgamated in 1893, though a connection to the Lowestoft line was built in the early 20th century. Holden D16/3 Class no. 62517 has a local train at Yarmouth Beach in the 1950s. Following closure, the station site became a coach park, with the canopies remaining until the 1980s when a complete reconstruction occurred. Photograph courtesy Rail-Online.

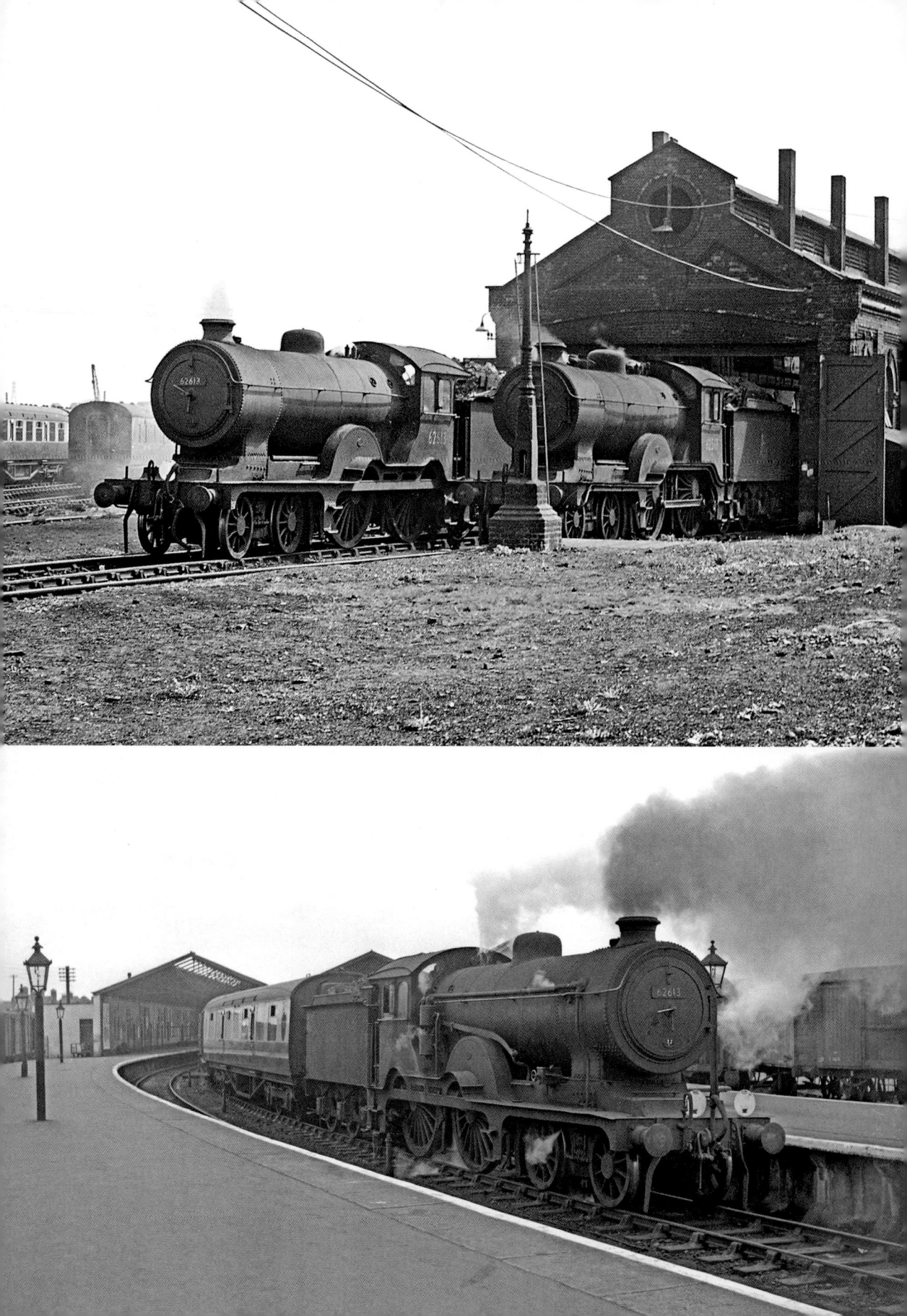

Opposite above **YARMOUTH VAUXHALL SHED**
Of the three sheds serving the three stations in Yarmouth, Vauxhall shed had the smallest complement of engines. Whilst Beach and South Town had around 20, Vauxhall was home to just two or three. The building was constructed on the west side of the station in 1883, replacing a structure from 1844 erected by the Yarmouth & Norwich Railway. Two Holden D16/3 locomotives protrude from Vauxhall shed on 26th May 1956. On the left is no. 62613 of South Town, whilst no. 62517 was on home ground. The depot was closed by BR in 1959. Photograph from the Norman Preedy Collection courtesy Rail Photoprints.

Opposite below **YARMOUTH VAUXHALL STATION – NO. 62613**
Yarmouth Vauxhall station was the first to be built, serving the line to Norwich from 1844. Whilst Yarmouth Beach and South Town stations opened subsequently, both have been closed. Vauxhall was rebuilt in 1960 following wartime damage and renamed Great Yarmouth station in 1989. Holden D16/3 Class no. 62613 has a Norwich train ready for departure from Vauxhall on 18th July 1955. Photograph courtesy Rail-Online.

Below **YARMOUTH SOUTH TOWN STATION – NO. 61233**
The 'Easterling' express ran for eight years, beginning 1950, between Liverpool Street, Lowestoft and Yarmouth South Town. Thompson B1 no. 61233 has just arrived at the latter with the train during 1956. Photograph courtesy Rail Photoprints.

BIBLIOGRAPHY

Allen, C.J. *The Great Eastern Railway*. 1968.
Allen, C.J. *Titled Trains of Great Britain*. 1983.
Banks, Steve and Clive Carter. *LNER Passenger Trains and Formations 1923-1967: The Principal Services*. 2013.
Bonavia, Michael R. *A History of the LNER 2: The Age of the Streamliners, 1934-39*. 1985.
Griffiths, Roger and Paul Smith. *The Directory of British Engine Sheds and Principal Locomotive Servicing Points: 1 Southern England, the Midlands, East Anglia and Wales*. 1999.
Groves, N. *Great Northern Locomotive History Volume 3a: 1896-1911 The Ivatt Era*. 1990.
Hooper, J. *The WD 'Austerity' 2-8-0: The BR Record*. 2010.
Hunt, David, John Jennison, Fred James and Bob Essery. *LMS Locomotive Profiles: No. 8 – The Class 8F 2-8-0s*. 2005.
Quick, Michael. *Railway Passenger Stations in Great Britain: A Chronology*. 2009.
RCTS. *A Detailed History of British Railways Standard Steam Locomotives: Volume One Background to Standardisation and the Pacific Classes*. 2007.
RCTS. *A Detailed History of British Railways Standard Steam Locomotives: Volume Two: The 4-6-0 and 2-6-0 Classes*. 2003.
RCTS. *A Detailed History of British Railways Standard Steam Locomotives: Volume Three: The Tank Engine Classes*. 2007.
RCTS. *A Detailed History of British Railways Standard Steam Locomotives: Volume Four: The 9F 2-10-0 Class*. 2008.
RCTS. *Locomotives of the LNER - Parts 1 to 10A*.
Sixsmith, Ian. *The Book of the Ivatt 4MTs: LM Class 4 2-6-0s*. 2012.
Walmsley, Tony. *Shed by Shed Part Two: Eastern*. 2010.
Wrottesley, John. *The Great Northern Railway: Volumes 1-3*. 1979-1981.
Yeadon, W.B. *Yeadon's Register of LNER Locomotives Volume Three: Raven, Thompson & Peppercorn Pacifics*. 2001.
Yeadon, W.B. *Yeadon's Register of LNER Locomotives Volume Four: Gresley V2 and V4 Classes*. 2001.
Yeadon, W.B. *Yeadon's Register of LNER Locomotives Volume Five: Gresley B17 & Thompson B2 Classes*. 1993.
Yeadon, W.B. *Yeadon's Register of LNER Locomotives Volume Six: Thompson B1*. 2001.

Also available from Great Northern

The Last Years of Yorkshire Steam
The Golden Age of Yorkshire Railways
Gresley's A3s
Peppercorn's Pacifics
London Midland Steam 1948-1966
The Last Years of North East Steam
British Railways Standard Pacifics
Western Steam 1948-1966
The Last Years of North West Steam
Gresley's V2s
Southern Steam 1948-1967
Yorkshire Steam 1948-1967

Gresley's A4s
Gresley's B17s
The Last Years of West Midlands Steam
East Midlands Steam 1950-1966
Thompson's B1s
The Glorious Years of the LNER
Scottish Steam 1948-1967
North East Steam 1948-1968
The Last Years of London Steam
Gresley's D49s
The Glorious Years of the GNR

visit www.*greatnorthernbooks.co.uk* for details.